Waste Matters

How do those pushed to the margins survive in contemporary cities? What role do they play in today's increasingly complex urban ecosystems? Faced with stark disparities in human and environmental wellbeing, what form might more equitable cities take?

Waste Matters argues that contemporary literature and film offer an insightful and timely response to these questions through their formal and thematic revaluation of urban waste. In their creation of a new urban imaginary which centres on discarded things, degraded places and devalued people, authors and artists such as Patrick Chamoiseau, Chris Abani, Dinaw Mengestu, Suketu Mehta and Vik Muniz suggest opportunities for an inclusive urban politics that demands systematic analysis. *Waste Matters* assesses the utopian promise and pragmatic limitations of their as yet under-examined work in light of today's pressing urban challenges.

This book will be of great interest to scholars and students of English Literature, Postcolonial Studies, Urban Studies, Environmental Humanities and Film Studies.

Sarah K. Harrison completed a PhD in English at the University of Wisconsin-Madison, USA.

Routledge Environmental Humanities
Series editors: Iain McCalman and Libby Robin

A full list of titles in this series is available at: www.routledge.com/series/REH

The *Routledge Environmental Humanities* series is an original and inspiring venture recognising that today's world agricultural and water crises, ocean pollution and resource depletion, global warming from greenhouse gases, urban sprawl, overpopulation, food insecurity and environmental justice are all *crises of culture*.

The reality of understanding and finding adaptive solutions to our present and future environmental challenges has shifted the epicenter of environmental studies away from an exclusively scientific and technological framework to one that depends on the human-focused disciplines and ideas of the humanities and allied social sciences.

We thus welcome book proposals from all humanities and social sciences disciplines for an inclusive and interdisciplinary series. We favour manuscripts aimed at an international readership and written in a lively and accessible style. The readership comprises scholars and students from the humanities and social sciences and thoughtful readers concerned about the human dimensions of environmental change.

Waste Matters
Urban margins in contemporary literature

Sarah K. Harrison

Routledge
Taylor & Francis Group
LONDON AND NEW YORK

earthscan
from Routledge

First published 2017
by Routledge

2 Park Square, Milton Park, Abingdon, Oxfordshire OX14 4RN
52 Vanderbilt Avenue, New York, NY 10017

Routledge is an imprint of the Taylor & Francis Group, an informa business

First issued in paperback 2019

British Library Cataloguing-in-Publication Data
A catalogue record for this book is available from the British Library

Library of Congress Cataloging-in-Publication Data
A catalog record for this book has been requested

ISBN: 978-1-138-18706-1 (hbk)
ISBN: 978-0-367-27122-0 (pbk)

Typeset in Times New Roman
by HWA Text and Data Management, London

Contents

Figures

Acknowledgments

I would like to thank my friends and colleagues at the University of Wisconsin–Madison where this book first took shape. Tejumola Olaniyan provided me with guidance and insight, and I am grateful for his continued mentorship. I am also grateful to Rob Nixon for his generous support of this project, and Nirvana Tanoukhi and Susan Stanford Friedman for their help and advice.

Thank you to Gwenola Caradec, Emily Clark, Krista Kauffmann, Mike Opest and Christa Tiernan for their friendship and support, especially when it was most needed.

Special thanks must go to my family, especially Mum, Dad and Ross who have supported me in countless ways, but always with love and laughter. Lastly, I must thank Denis, for encouragement, patience and perspective; and Faye, for joyful distraction. You both make nothing else, and everything else, matter.

An earlier version of Chapter 2 appeared in *Research in African Literatures* 43.2 (2012). I am grateful to Indiana University Press for permission to reprint that material here.

Introduction

Locating urban waste

The bodies keep coming. First, the corpse of a teenage boy floats to the surface of Accra's polluted Korle Lagoon, then another is found in a muddy ditch in the city's historic Jamestown district. A young woman is thrown onto a garbage dump. Next, a dead boy is left in a filthy public latrine in one of the city's busy lorry parks. In *Children of the Street*, Kwei Quartey's Inspector Darko Dawson pursues a serial killer whose signature disposal of his young, homeless victims 'express[es] that these people's lives are worthless to him. They might as well be rubbish or refuse' (Quartey 2011, p. 256). The perpetrator is revealed to be Obi, himself a former street child who is now an 'all-round handyman' for the renowned psychologist Professor Allen Botswe (Quartey 2011, p. 115). Charged with bringing children home for his employer to interview about how they cope with life on the streets, Obi suffers his own tragic and violent breakdown. After his arrest, he indignantly tells Dawson:

> He brings this filth, this refuse from the streets to sleep in the spotless sheets I wash with my own hands. No. Those children do not belong here. They belong in the gutter or the latrine. That is all they are worth.
>
> (Quartey 2011, p. 321)

Having narrowly escaped his own life of poverty by finding work in Botswe's mansion, the traumatised Obi, Dawson muses, 'may be trying to kill that part of [the street children] that's in him' (Quartey 2011, p. 257).

Filled with detailed descriptions of Ghana's busy capital and its diverse residents, Quartey's gripping narrative vividly dramatises the material and imaginative challenges produced by contemporary urbanisation. While Obi's psychopathy is a deliberately heightened example, his violent disposal of his victims is tragically indicative of the widespread social disregard for many of Accra's most vulnerable residents. Attracted to the city by the promise of economic opportunity, this supply of largely unskilled workers outstrips the demand for their labour. As they accumulate at the literal and figurative urban margins, they are exposed to the poverty, pollution, overcrowding and crime that accompany intense urban growth. As Obi tells an indignant Inspector Dawson, 'Street people are sleeping everywhere. Who knows they are there, and who cares about them? Who will report anything?' (Quartey 2011, p. 322). Physically ubiquitous, yet socially ignored, these disempowered urban residents are literally reduced to the status of the detritus in which they frequently live and work.

Like Obi, Dawson struggles to reconcile the extremes of poverty and wealth that characterise today's developing cities. Far more measured in his response, he is nevertheless disturbed by the contrast between Professor Botswe's opulent mansion and the apocalyptic landscape of Agbogbloshie, 'Accra's most notorious slum', where many of the novel's eponymous street children end up working (Quartey 2011, p. 9). Here, youngsters form an inverted and risky '*dis*assembly' line, dismantling 'junked, unusable equipment that rich countries pass off as charitable donations' to retrieve copper that can be resold (Quartey 2011, pp. 9–10). Together with the surrounding landscape, they are exposed to highly toxic chemicals in the course of doing so. A false moral economy underpins the export of this obsolescent technology from the global North to the African continent. Characterised by its more economically powerful 'benefactors' as a benign form of recycling that reduces e-waste and provides for those in need, this global waste flow in fact brings with it severe physical and environmental costs for its recipients. The progressive premise of capitalist production is literally dismantled in Agbogbloshie, where the informal e-waste economy toxically inverts the manufacture of such items as mobile phones, computers and electronic cables. Quartey's description of this perilous setting highlights the inseparability of economic development from the production of not only material waste, but also discarded humans. Sociologist Zygmunt Bauman (2004) has critiqued this unwelcome symbiosis, arguing that globalisation is primarily an economic phenomenon energised by a destructive capitalism that relentlessly produces large numbers of 'wasted lives' in its quest for economic progress at the expense of equitable, humane modes of production. Yet what serves for Bauman as a useful critical analogy is offered as a disturbingly accurate description by Quartey's novel. In Agbogbloshie, the metaphorical distance between waste matter and wasted humans has diminished to the point of invisibility.

The urban deprivations so finely rendered by Quartey's fiction proliferate throughout the cities of the developing world. Urbanist Mike Davis provides a damning account of this widespread urban poverty in his compelling comparative analysis of what he terms our 'planet of slums', in which he notes, recalling Bauman, that 'the principal function of the Third World urban edge remains as a human dump' (Davis 2005, p. 47). Some have objected to Davis's broad application of the term 'slum' to diverse locations within the global South, given its origins in middle-class Victorian discourses of urban moral decline that deployed imperial tropes to construct Britain's inner-city poor as 'savage' social deviants.[1] In today's cities, Tom Angotti argues, the term still obscures the structural reasons for disproportionate urban poverty, naturalising the decline of minority neighbourhoods, and legitimating 'high-minded' urban renewal (Angotti 2006, p. 961). However, to label slums otherwise risks euphemising precisely the unease that they do, and should, provoke. While Davis offers an urgent critique of their present status accompanied by alarming projections into their future, his insistence on the term 'slum' also productively suggests the long, intertwined histories of these formations. David Cunningham is, in fact, persuasive in his suggestion that 'if the term retains a productive force today, in the context of a globalising capitalism, it is precisely – through a recollection of its roots in the nineteenth-century metropolis – in the degree to which it recalls the distinctive modernity of the social–spatial forms it now so riskily names' (Cunningham 2007, p. 16). There is a significant continuity between the intense urban poverty that once undermined imperial London's claim to industrial progress and that which has been displaced by newly distantiated modes of production to the urban margins of formerly colonised cities. This is not to suggest, of course, that Britain's urban poverty has been wholly outsourced. However, the particular intensity of contemporary slum formation in the global South must be understood in the context of its long colonial history, which now sees more economically powerful countries disingenuously regrouping under the sign of globalisation.

While contemporary imperial capitalism is one of the engines of contemporary slums, they are not solely economic formations. Often, but not always located at the geographical edges of developing cities, these sites are distinctive marginal ecologies where inseparable material, social and environmental inequities are seen and felt with particular intensity. Asserting their exemplary nature, Pablo Mukherjee identifies life at the urban margins of the developing world 'where, to a virtually unprecedented degree . . . the historical condition of unevenness is felt and lived as a toxic environmental condition – as the condition of postcoloniality itself' (Mukherjee 2010, p. 90). Physically speaking, these spaces are profoundly unstable. Whether constructed on swampland, like the Maroko slum in Lagos that Chris Abani depicts in *GraceLand* (2004), or on the unstable, toxic ground of vast garbage dumps such as those found in cities as distant as Rio de Janeiro and Cairo, today's slums are in-between places, whose landscape is shifting together with their identity and purpose.

Waste Matters seeks to better understand these new and unsettling city spaces by examining a range of literary and visual texts – novels, non-fiction narratives, films, photographs, visual artworks – that question their origins, describe their contours and imagine their alternatives. The fundamental premise of the book is that contemporary representations of these instructive sites offer valuable tools with which to interrogate the means by which urban inequality is produced and sustained. Focusing on work by Patrick Chamoiseau, Chris Abani, Dinaw Mengestu, Suketu Mehta and Vik Muniz, *Waste Matters* shows how such artists explore opportunities for social and environmental justice through their particular engagement with the politics and aesthetics of *urban waste* – a necessarily expansive concept that designates not only discarded things and degraded places, but also the devalued people that feature as both symptoms and symbols of postcolonial inequity in the texts examined. In addition to thematically revaluing urban waste by placing marginal figures and spaces at the centre of their narratives, these authors and artists appropriate forms of waste management such as gleaning and recycling in the composition of their texts. While early postcolonial literature by writers such as Ngugi wa Thiong'o, Wole Soyinka and Ayi Kwei Armah expresses notable urban disillusionment, *Waste Matters* finds that contemporary postcolonial artists identify the margins of today's cities as potent sites for the definition and rehearsal of new modes of citizenship and belonging that eschew restrictive national and ideological frameworks. However, as the explicit violence of Quartey's novel suggests, they also avoid naïve optimism in favour of carefully describing and deliberately problematising the possible recuperation of the urban margins.

Critical contexts

In its attention to the varied forms of urban waste that feature prominently in contemporary texts, *Waste Matters* intervenes in the thriving interdisciplinary field of Waste Studies, which seeks to clarify the multivalent significance of rubbish, filth, toxins and remains across a wide range of contexts.[2] The broad scope of such an inquiry allows for productive conceptual variation. Sarah A. Moore notes that 'the preponderance of research on waste' in geography identifies it 'as having a specific characteristic that defines it and as something that is largely external to society' (Moore 2012, p. 783). Literary criticism is well placed to complement this positivist view by showing how waste is figured as both a physical problem and unwelcome social status in a range of literature. An inescapably socio-material formation, waste's literal dimensions are inseparable from its figurative import.

To date, literary analyses of waste have largely focused on American texts.[3] This rich and varied work commonly explores how artistic engagements with trash, toxins, dirt and excrement offer suggestive counterpoints to the social exclusions and environmental degradations routinely generated by a conspicuously consumptive Western capitalism. John Blair Gamber's *Positive Pollutions and Cultural Toxins* (2012) provides an especially resonant analysis of the ways in which contemporary US ethnic literatures affirm the value of wasted city spaces in light of prevalent racialised discourses of American urban 'doom'. Chapter 3 reprises this concern with American urban margins by analysing how the exilic perspective of Dinaw Mengestu's Ethiopian protagonist in *The Beautiful Things That Heaven Bears* (2007) affords an original critique of urban transformation in Washington, D.C.

These concerns resurface with particular intensity in the postcolonial context. Sarah Lincoln's *Expensive Shit* (2008) and Kenneth Harrow's *Trash: African Cinema from Below* (2013) relatedly argue that the prevalent trope of waste in African literature and film offers an effective aesthetic revaluation of the people and cultural productions degraded by the continent's uneven imbrication with global capitalism.[4] While some might be wary of attributing 'trashiness' to African art, *Waste Matters* endorses this multivalent heuristic.[5] Far from naturalising the devaluation of postcolonial literature or its subjects, critical attention to urban waste in its many forms enables a better understanding of that precarious status and the innovative ways through which it is contested and overcome.[6]

In addition to broadening the geographic scope of these existing analyses, *Waste Matters* contributes to this ongoing critical conversation in several ways.

First, as a crucial interface between humans and the environment, a focus on waste in contemporary urban literature enables further elaboration of a literary critical perspective that productively combines postcolonial and ecocritical concerns. In an influential analysis of the two fields, Rob Nixon (2005) notes their historic tendency towards divergence, if not outright opposition. While once dominant environmentalisms had been preoccupied with the study and preservation of specific bioregions, postcolonialism was traditionally more invested in anthropocentric historical and cultural critique. Although a large number of critics have since put these approaches into useful conversation,[7] surprisingly few have turned their attention to postcolonial waste, a particularly notable omission given its distinctive contemporary urban formation.[8]

Acknowledging the continuum of concerns that link people and planet has significant implications for postcolonial studies as Ann Laura Stoler suggests when she urges scholars in the field to 'refocus on the connective tissue that continues to bind human potentials to degraded environments, and degraded personhoods to the material refuse of imperial projects' (Stoler 2008, p. 193). In highlighting the linked concerns of postcolonial landscapes and populations, Stoler builds on the valuable work of environmental historians such as Alfred Crosby (1986) and Richard Grove (1995) that has pointed out the ecological imperatives at the heart of European empire-building. If, as these scholars suggest, the history of imperialism is one of shared human and environmental degradation, a postcolonial studies committed to revealing both its legacies and lived effects must be attentive to this imbrication.

Likewise, a key insight of much ecocriticism has been to highlight the inseparability of anthropological and environmental concerns. In its call for the agentic capacities of all matter to be taken seriously, scholarship under the rubric of the 'new materialisms' dismantles a false human/nature dichotomy that suggests the primacy of the former.[9] Such work moves beyond what Anthony Lioi identifies as the 'dirt-rejecting' tendency of earlier ecocriticism to acknowledge the messy, 'impure' materiality shared by all bodies, things and places (Lioi 2007, p. 17). The production, disposal and management of waste clearly plays a key role in understanding this connectedness. Our physical ingestion and excretion of different kinds of waste – toxins, pollutants, shit – signals the mutual permeability of our human bodies and a supposedly distinct natural world from which we are inextricable. However, the ways in which we organise discarded matter – trash, leftovers, remnants – reveal attitudes towards the environment that remain inherently imperial.

Waste Matters builds on such scholarship in its collective definition of urban waste as things, places and people that have commonly been discarded. However, while acknowledging the ontological connection of these diverse agents, this book resists their categorical equation. In particular, the troubling conflation of marginal humans with physical waste is interrogated throughout. Urban waste's material potency is described and dramatised in all the texts examined, but this is always inseparable from its sociocultural constitution. *Waste Matters* reveals how and why waste comes to be defined as such in postcolonial cities, finding in this complex assemblage the basis for imagining and demanding social and environmental justice.

Discourses of waste

The rapidity with which urban waste has proliferated in recent decades, coupled with the immediacy of the challenges it presents, lends, in some ways, a distinctive contemporaneity to these problems. However, they are by no means exclusively recent phenomena. As Martin Melosi explains in the comprehensive introduction to his *Garbage in the Cities*, 'since human beings have inhabited the earth, they have generated, produced, manufactured, excreted, secreted, discarded, and otherwise disposed of all manner of waste' (Melosi 2005, p. 1). In turn, literature and the arts have long offered a means of describing and analysing waste's prevalence. Eleanor Johnson, for example, argues that at least since the late Middle Ages, artists have 'marshal[ed] poetic resources as means for synthesising and forming an ideology of waste and its consequences for human society' – a concern that continues to preoccupy the contemporary writers examined in *Waste Matters* (Johnson 2012, p. 473).[10]

The longevity of the urban waste problem should not, however, be mistaken for its timelessness. Melosi identifies nineteenth-century industrialism as a period in which refuse proliferated with particular intensity in American cities. Strategies for its management were tied to what he describes as 'a unique set of circumstances', including shifting notions of sanitation, municipal responsibility and civic engagement (Melosi 2005, p. 2). As his historical account reveals, definitions of and attitudes towards urban waste change over time, contingent on specific social factors.

Like Melosi, anthropologist Mary Douglas also suggests that the significance of waste is socially constructed. Yet while Melosi focuses on the excessive accumulation of solid refuse in urban environments, Douglas compares 'concepts of pollution and taboo' in a range of settings, asserting that the ritual production of 'dirt' is used to shape and consolidate social structures. As she explains, 'where there is dirt there is system. Dirt is the by-product of a systematic ordering and classification of matter, in so far as ordering involves rejecting inappropriate elements' (Douglas 2005, p. 35). While she has been criticised by some for the relative conservatism of her ideas, her assertion that 'pollution behaviors' are deliberately marshalled to effect social order usefully suggests how contemporary waste practices are used to strengthen particular socio-economic hierarchies.[11] While contemporary urban waste differs in degree and in kind from the 'dirt' which preoccupied Douglas, the exclusionary measures that shape today's cities – from immigration laws to urban redevelopment – echo precisely the ordering impulse that she examined. Mired in discourses of waste that call their societal contributions into question, the status of today's slum-dwellers is paradoxically precarious and immobile.

Both Melosi and Douglas offer useful precedents for exploring the distinctive discourses of waste that emerge during colonialism. Across a range of contexts, historians have identified how colonisers systematically conflate indigenous populations with unwanted matter in order to help maintain their authority. By routinely pathologising supposedly 'filthy natives', imperial racism is bolstered by sanitary paranoia. Thomas R. Metcalf's *Ideologies of the Raj* offers a useful analysis of how, for example, 'India's disease and dirt became markers of its enduring "difference", and so helped sustain the larger ideology that undergirded the Raj' (Metcalf 1995, p. 173). In his related analysis of early twentieth-century American colonialism in the Philippines, Warwick Anderson similarly demonstrates the manner in which American medical and scientific literature systematically constructed Filipinos as 'open, threatening, excreting animals' in dire need of colonial 'civilisation' (Anderson 1995, p. 651). Despite their historical and geographical particularities, representations of waste thus emerge as a central mechanism of social control common to these, and other, colonial situations.[12]

Julia Kristeva's *Powers of Horror* (1982) suggests how the discursive construction of abject colonised populations performs an important psychological function for those ostensibly in power. Drawing on Douglas' work, Kristeva asserts that abjection is not a quality that inheres in a given object, but rather the crisis of self-identity produced through encounters with 'what disturbs identity, system, order. What does not respect borders, positions, rules. The in-between, the ambiguous, the composite' (Kristeva 1982, p. 4). Taking the corpse as her primary example, Kristeva argues that evidence of bodily decay, including pus, shit and blood, calls into question our physical integrity, troubling both the actual and psychological borders of the self. Colonisers attempt to consolidate identity and authority through their contradictory separation from that which they ambiguously designate as waste, yet, whether material refuse, marginalised people, or unused spaces, they cannot completely distinguish themselves.

The colonial city

Colonial waste practices are put under particular pressure by the paradoxical nature of urban occupation. Although segregation ostensibly demarcates those in power from those subjected to it, interaction is both inevitable and encouraged by proximity, employment, desire and resistance to the status quo. Herein lies what might, following Homi Bhabha, be termed the 'ambivalence' of the colonial city – an untenable desire for demarcation, which asserts the distinctiveness of colonial practices, ideals and sensibilities, while simultaneously holding out the possibility of their subjects' assimilation into that same social order.[13] If, as Douglas reminds us, dirt is 'matter out of place', colonial cities are sites of anxious displacement, in which native populations are violently uprooted and oppressively confined, sometimes simultaneously (Douglas 2005, p. 35). Sympathetic to the complex motivations of the societies that she examines, Douglas finds that 'eliminating [dirt] is not a negative movement, but a positive effort to organise the environment' (2005, p. 2). In the colonial context, however, such spatial organisation most often signifies a systematic disregard for existing inhabitants and a condescending assumption of best environmental practice.[14]

In *The Wretched of the Earth* (2004 [1961]), Frantz Fanon memorably describes the palpable tension produced by colonial efforts to maintain the spatial organisation necessary to the consolidation of their authority. The oppositional colonial city that he describes is, following Henri Lefebvre (1991), a socially produced space, compartmentalised into mutually exclusive 'native' and 'colonists'' sectors. Police officers and, in some cases, soldiers, forcibly maintain the division between the two, demonstrating that the colonial imposition of what Lefebvre calls 'representations of space' – the abstract space conceived by urban planners, cartographers and engineers – cannot be achieved without implied or actual violence (Lefebvre 1991, p. 38). Indeed, as Fanon explains, it is precisely this threat which the colonised internalise and eventually repeat in their aggressive resistance to colonisation. The colonial production of urban space thus fails to account for its inherent multi-dimensionality. As Lefebvre explains, spatial experience, or 'lived space,' emerges from the interaction between 'conceived spaces' and 'perceived' spaces; in other words, between hypothetical spaces we imagine and the observable and concrete world, the material world (Lefebvre 1991, p. 39). Spatial abstractions cannot be realised without concessions to the plural realities of urban existence. Wasted humans – exiles, refugees, those forced into homelessness or 'resettlement' – are the inevitable byproduct of any attempt to literalise abstract space. From a colonial viewpoint, they are the 'collateral damage', the excess that must necessarily be eliminated in order to reach a desired ideal.

The colonised's sector that Fanon describes preempts the urban margins of today:

> [T]he 'native' quarters, the shanty town, the Medina, the reservation, is a disreputable place inhabited by disreputable people. . . . It's a world with no space, people are piled one on top of the other, the shacks squeezed tightly together. The colonised's sector is a famished sector, hungry for bread, meat, shoes, coal, and light. The colonised's sector is a sector that crouches and cowers, a sector on its knees, a sector that is prostrate.
>
> (Fanon 2004, p. 4)

Overcrowded, lacking material resources and adequate provisions, this is a place where the most basic human needs are unmet. The colonist's sector, by contrast, is:

> a sector built to last, all stone and steel. It's a sector of lights and paved roads where all the trash cans constantly overflow with strange and wonderful garbage, undreamed-of leftovers. . . . The colonist's sector is a white folks' sector, a sector of foreigners.
>
> (Fanon 2004, p. 4)

This is a space of plenty, of excess, whose inhabitants can readily satiate their desires. Of particular interest is the abundant and enticing rubbish: to produce waste is to possess power; it is the inability to separate oneself from waste that is fatal. Those who inhabit the 'native' sector, like many who reside in contemporary slums, can barely manage to feed themselves, let alone afford the conspicuous consumption that is the distinguishing feature of urban affluence. For the colonists, waste is not wholly undesirable; whether human, environmental, or material, it reinforces the actual and figurative boundaries between their urban comfort and the harmful discrimination on which it depends.

Although Fanon convincingly depicts the untenability of the colonial status quo in *The Wretched of the Earth*, he is equally critical of the new national bourgeoisie whose acquisitive self-interest, he presciently argues, will quickly become apparent after decolonisation. Dismayed by their unwillingness and inability to effect the economic overhaul that would sever neocolonial ties, Fanon notes the urban bias of the new ruling class, which prevents them from meeting the needs of the entire country, especially those in rural areas.

The apparently pestilential cities that Fanon envisages as 'teeming with the entire managerial class' after decolonisation (2004, p. 122) preempt the urban disillusionment of mid-twentieth-century postcolonial fiction by authors such as Wole Soyinka and Ayi Kwei Armah. As Joshua D. Esty notes in 'Excremental Postcolonialism', such authors deploy scatological metaphors to condemn the failed promises of a national elite that prioritises personal wealth instead of collective progress. In *The Beautyful Ones Are Not Yet Born* (Armah 1968), for example, Armah's depiction of a corrupt government minister's ultimately self-abasing desire for wealth 'reodorises money, converting it into shit and forcing readers to see wealth as polished waste' (Esty 2007, p. 33). No longer 'strange and wonderful', as Fanon ironically noted, the excesses of the Europeanised comprador class are figured as actually and morally contaminated (Fanon 2004, p. 4).

The excremental counterdiscourse that Esty reveals is not only directed at the neocolonial elite, however. He argues that, beyond satirising political corruption, the trope of shit performs 'an autocritical function' for postcolonial writers (Esty 2007, p. 36). In works by Armah and Soyinka, together with James Joyce, their protagonists' recoil from the literal and figurative filth of public engagement dramatises the writers' own struggle to reconcile 'ethical selfhood and aesthetic freedom' with 'the burden of national representation' (Esty 2007, p. 55). In these texts, bodily waste thus serves as an important indicator of the tension between social critique and social withdrawal that is inherent in postcolonial writing.

Global contact zones

Whereas the literature that Esty examines expresses an urban disillusionment that is symptomatic of authorial unease with a broader national collectivity, contemporary postcolonial writing foregrounds urban waste in order to explore concerns about identity, citizenship and belonging in an even more globalised world. As James Holston and Arjun Appadurai point out, 'cities are challenging, diverging from, and even replacing nations as the important space of citizenship – as the lived space not only of its uncertainties but also of its emergent forms' (Holston and Appadurai 1996, p. 189). This is not to suggest that the national context is no longer significant, far from it – in fact, it is in part the tension between local, urban, national and transnational affiliations that lends the cities examined here their particular energy.

The divided colonial city that Fanon so memorably described no longer exists, if it ever truly did. Whereas his urban portrait strategically subdued the intermingling of different urban populations in favour of emphasising the oppositional tension of colonial urbanisation, intermixture is, in fact, a longstanding characteristic of metropolitan life. Urban growth, infrastructural collapse, shifting employment patterns and cultural events bring about a diverse range of encounters between otherwise disparate urban residents. Today's cities are, following Mary Louise Pratt, distinctive 'contact zones' in which the 'highly asymmetrical relations of domination and subordination' that mediate such processes of urban exchange are more pronounced than ever before (Pratt 1992, p. 4).

While formal decolonisation has resulted in the scaling back of many overt mechanisms of urban surveillance and control, imperial traces endure in the multivalent architecture of such cities as Mumbai, where prominent public buildings such as Victoria Terminus evoke British efforts to imprint colonial authority on the very fabric of the city. Renamed Chhatrapati Shivaji Terminus after a seventeenth-century Indian king in 1996, two Islamist militants fatally shot 58 people inside this iconic station during the 2008 Mumbai terrorist attacks. At once the site of erstwhile colonial prestige and the arena for new forms of urban militarism arising from new international conflicts, the station popularly known as 'VT' marks the emergence of new, complex divisions with the contemporary postcolonial city. The historical asymmetry between European colonisers and Indian city-dwellers is now overlaid by new geographies of exclusion and violence fostered by the identarian politics that have taken hold in India since the 1980s. Such conflicts do not only fall along international axes, but also produce new, shifting divisions within the cities themselves. As discussed in Chapter 4, the city had already been splintered along complex class and ethnic lines by the Hindu–Muslim riots of 1992–93, which preceded the 2008 violence. Fanon's Manichean city is now profoundly fragmented with residents' affiliations forming and reforming in often unpredictable fashion.

As urban margins shift and proliferate, contemporary urban scholars must seek new ways of registering this dynamism in their analyses. Guido Martinotti tellingly laments 'the image of the dormant city' produced by the empirical data on which many urbanists rely (Martinotti 1999, p. 178). If, as he suggests, 'direct observation tells only a partial story about urban society', postcolonial literature complements this narrative by describing, dramatising and reimagining the lived experience of contemporary cities (Martinotti 1999, p. 177). The texts examined throughout movingly individualise the city dwellers who are collectively analysed by urban scholars.

Furthermore, as James Donald notes, literature plays a constitutive role in urban life, with cities best understood as complex 'imagined environment[s]' composed of the layered narratives told by their various residents, visitors, and observers (Donald 1999, p. 8). Çinar and Bender likewise assert that cities are inherently 'imagined places' (Çinar and Bender 2007, p. xii). Bringing literary insights to bear on urban analyses thus allows for proper consideration of the significant imaginative component of city existence.

While strategic generalisation enables urban scholars to usefully delineate common patterns or trends within and between cities – the diachronic emergence of particular migrant trajectories, for example – in the colonial context, this descriptive mode has historically been used as a means of asserting relative distinction, and therefore authority. In addition to lending affective texture to urban studies, contemporary literature thus has an important counter-discursive role in rebutting colonial stereotypes and contesting new ones. John Scanlan suggests that Western epistemology relies on the undifferentiated representation of marginal, disenfranchised and minority groups, a process he equates to metaphorical waste-making. As he explains,

> Garbage indicates the removal of qualities (characteristics, or distinguishing features) and signals the return of everything to some universal condition, perhaps impersonal... At a human level a violent stripping away of (positive) characteristics consigns its victims to an indistinguishable mass, a state that ensures their treatment as mere rubbish – social outcasts, foreigners, others ... simply stuff that can be pushed around, co-mingled with its similarly valueless and indistinguishable like.
>
> (Scanlan 2005, p. 34)

In *An Area of Darkness*, V. S. Naipaul expresses the psychic vertigo that results from such unwelcome anonymity. Shaped by his Caribbean upbringing and European education, he is unsettled by his lack of ethnic distinction during his first visit to his ancestral homeland of India:

> For the first time in my life I was one of the crowd. . . . Now in Bombay I entered a shop or a restaurant and awaited a special quality of response. And there was nothing. It was like being denied part of my reality. Again and again I was caught. I was faceless. I might sink without a trace into that Indian crowd. I had been made by Trinidad and England; recognition of my difference was necessary to me. I felt the need to impose myself, and didn't know how.
>
> (Naipaul 1964, p. 46)

Consigned to the 'indistinguishable mass' that Scanlan posits as the necessary Other to Western self-identity – a theoretical move that recalls Edward Said's *Orientalism* (1978) – Naipaul's metropolitan affiliation is unsettled. This passage reveals not only the subjective uncertainty wrought by Naipaul's imperial sympathies, but also suggests the existential crisis faced by those who must routinely struggle against the anonymity and thus worthlessness imposed by the state. Whereas, as shown in Chapter 4, Naipaul ultimately maintains a studied distance from Bombay's many urban poor, Suketu Mehta relates multiple interlinked biographical stories from Bombay's urban margins in *Maximum City* (2004). In doing so, he constructs a compelling counterpoint to the discursive 'wasting' of slum-dwellers and the poor by articulating the real, unacknowledged rights of those consigned to slums, ghettos and degraded neighbourhoods.

Although they are responding to longstanding forms of prejudice and discrimination, contemporary urban texts such as those examined here are not simply reactionary. Mehta's polyphonic narrative does not offer an easy celebration of Bombay's margins, instead engaging with a range of the city's poor, including gangsters who commit disturbing violent crimes, not bold revolutionary acts. In its fantastical rendering of a near-future Johannesburg, Lauren Beukes' impressive *Zoo City* (2010) goes even further to accentuate the flaws of those who reside in the city's eponymous criminal ghetto. Convicted of various misdeeds, these explicitly marginalised urban dwellers are burdened with unshakable animal familiars that serve as the constant, physical embodiment of their guilt. Beukes draws attention to the ways in which these outcasts creatively adapt to their status as urban waste, making innovative use of its material forms. On exiting her rundown apartment block, her protagonist Zinzi observes the stairwell's missing floorboards, pipes and lightbulbs, wryly noting that this is 'the way of the slums. Even the stuff that's nailed down gets repurposed' (Beukes 2010, p. 12). In this case, however, the stolen items have been sold for heroin or refashioned into pipes for smoking meth. The growing engagement with urban waste that is evident in contemporary postcolonial literature is, as we see here, multi-dimensional – the authors examined invite nuanced readings of this trope, rather than simplistic praise for its recuperation.

The vandalised apartment block in which Beukes' Zinzi lives speaks to the tension between the individual experiences that such texts dramatise in all their complexity, and the need to forge a space for collective action. Excluded from, or in this case, severely restricted by dominant structures of power and authority, how can marginal urban residents retrieve a sense of civic responsibility? On what grounds might they form affiliations, allowing them to work together for a communal good? While delineating individual lives is a key role of postcolonial urban literature, imagining a space for social and political cooperation emerges as an equally important task for the texts examined. For Patrick Chamoiseau, as discussed in Chapter 1, this takes the form of a distinctive historiographical project in *Texaco* (1997), in which he asserts the shared past of both Fort-de-France's central and peripheral residents as a means to advocate the incorporation of the city's slum-dwellers into the city proper. In Chapter 2, Chris Abani's *GraceLand* (2004) is less optimistic about the possibility of civic unity in the face of historic and current inequities, dramatising his protagonist's eventual flight from the embattled urban margins of Lagos. What these texts share, however, together with Mengestu's and Mehta's work, is a commitment to the function of art in facilitating political consciousness and, by extension, necessary social change. Abani explicitly dramatises this when, shortly after performing with the Joking Jaguars music troupe, his protagonist Elvis experiences an epiphanic realisation of his vulnerable place as a young slum-dweller within an unevenly stacked global system. Music, as one of his bandmates remarks, has the double-edged ability to both conceal and reveal 'de knife-edge beauty of seeing yourself as you are. As you really are' (Abani 2004, p. 276). Likewise, the texts examined in *Waste Matters* foreground both the hostile reality that is the condition of life at the urban margins while also imagining its possible alternatives and solutions.

Reading transnationally

While local, grassroots activism is offered as one significant mode of articulating marginal rights in each of the texts examined, they gesture towards the need for collaboration on a larger scale. Their focal cities – Fort-de-France, Lagos, Washington D.C., Mumbai and Rio de Janeiro – exceed their immediate context. They are not only sites of specific imperial legacies and postcolonial dilemmas, but they are also globally connected through circuits of migration, business and technology. Moreover, as this literature is uniquely placed to explore, these connections are supplemented by intangible linkages of memory, desire, ambition, faith and prejudice, which extend each city's lived boundaries. Urban simultaneity emerges as a particular hallmark of life at the urban margins, where specific, localised experience is always concurrent with actual and imagined connections to other historically and geographically disparate cities.

The impressive globality of contemporary cities vividly demonstrates what David Harvey has called the 'time-space compression' of the postmodern age, the result of those 'processes that so revolutionise the objective qualities of space and time that we are forced to alter, sometimes in quite radical ways, how we represent the world to ourselves' (Harvey 1989, p. 240). However, the diverse texts examined here trouble his understanding of contemporary cities as increasingly homogeneous, characterless places. Despite their various connections and underlying structural similarities, postcolonial cities emerge as textured, idiosyncratic and dynamic. The representational challenge that time-space compression thus demands of these authors is that of articulating both the immediate need and also the global solidarity of the urban margins. Some NGOs are already facilitating transnational collaboration in face of the globally widespread problem of slums. Shack/Slum Dwellers International, and the Informal Waste Pickers and Recyclers Project, for example, both distribute resources to and centralise knowledge from a global network of marginal urban communities. In their creation of a new urban imaginary that couples aesthetic realism with measured utopianism, the texts examined similarly invite connections between marginal groups whose experiences, while distinctive, can no longer be understood in isolation from one another.

Waste Matters similarly negotiates these interrelated problems of scale and specificity. In noting the instructive commonalities of disparate literary sites, I am mindful of Kalpana Seshadri-Crooks' critique of 'an undifferentiated notion of the margin' that circulates within postcolonial studies to become 'fetishised and reified as the dislocated and authoritative *critical* position' (Sheshadri-Crooks 1995, p. 66). Wary of instrumentalising these texts in the service of a reading that assumes their uniform 'resistance' to a perceived urban norm, each chapter pairs comparative generalisations with close readings that are attentive to the specific historical contexts and formal strategies of each of the texts examined.

Each chapter focuses on narratives of a different city: Fort-de-France, Lagos, Washington, D.C., Mumbai and Rio de Janeiro. These texts are placed within their distinct national and literary traditions, but their situation within a single canon is necessarily troubled by the migrant experiences that they commonly evoke. The cities that they depict are, as mentioned above, infused with the material and ethereal traces of other places. Following Michael Peter Smith (2001), these narratives envisage, dramatise and interrogate 'transnational urbanism'. By elaborating on what Smith terms the 'criss-crossing transnational circuits of communication and cross-cutting local, translocal, and transnational social practices' that collide and intersect in these representations, the chapters invite connections between the various cities explored (Smith 2001, p. 5). More specifically, examining the relationships within and between the margins of these cities foregrounds what Françoise Lionnet and Shu-mei Shih (2005) term the 'minor transnationalisms' that are easily elided by comparative urban and literary studies. While the globalisation of which slums are a signal characteristic has been taken by some to spell the critical demise of postcolonial studies, which was once predominantly concerned with asserting the sovereignty and integrity of national cultures, *Waste Matters* reveals urban waste to be a crucial interface *between* the local and the global, where the desire for secure national citizenship competes with the lure of transnational social and economic mobility.[15]

This transnational reading practice not only complements the themes and forms of the texts examined, it also extends existing frameworks of urban comparison that have tended to divide the study of cities into hierarchically opposed 'First-' and 'Third Worlds'. While dominant paradigms of 'world' or 'global' cities strongly emphasise the degree to which different cities are integrated into the global economy,[16] *Waste Matters* as a whole responds to recent calls for more equitable terms of global urban comparison by putting narratives of 'developed' and 'developing' cities into direct conversation.[17] Yet whereas Jennifer Robinson (2006) proposes an 'ordinary-city' approach, which eschews First World exceptionalism in favour of bringing all cities into the same analytical plane, the subsequent chapters look to the unusual, unexpected and, in some cases, extraordinary potential of the urban margins that contemporary postcolonial artists evoke through their engagement with urban waste.

Book overview

Chapter 1 examines a novel that usefully situates the urgent daily struggles of contemporary slum-dwellers within a longer history of imperial domination and metropolitan influence. In *Texaco* (1997), Patrick Chamoiseau uniquely imagines the circumstances that have led to the formation of the eponymous slum on the outskirts of Fort-de-France, Martinique's capital city. While critical accounts of urban history typically focus on official discourse and formal architectural practices, Chamoiseau explores the material and discursive challenges faced by those who lack access to protected sites of remembering. Extending existing analyses of his historiographical project, *Waste Matters* argues that Chamoiseau develops a dynamic ecological narrative method to tell the story of the slum's emergence. Drawing on the aesthetics and thematics of urban gleaning, he narrates Martinican history in a manner that seeks to protect the island's cultural integrity in the face of its traumatic colonial past and troubled neocolonial present. The tension between the novel's form and content – on the one hand, its textual excess; on the other, its narrative of material lack – troubles the simplistic recuperation of that which has been discarded, devalued and degraded. In doing so, *Texaco* refuses to naturalise or legitimate the long-standing marginalisation of Martinique's slum-dwellers.

While *Texaco* concludes on a note of qualified optimism with official recognition of the slum and its residents, Chapter 2 examines a novel that portrays the urban margins in a more critical light. In *GraceLand* (2004), Chris Abani extends a long tradition of Lagos literature by re-imagining the city from the perspective of its poorest population. Set in the 'swamp city of Maroko' during the early 1980s when this inner-city slum was slated for destruction, the novel asks what, if anything, there is to be salvaged from this urban 'wasteland'. This chapter takes up this question by analysing the education that Abani's protagonist Elvis receives when he migrates to the then Nigerian capital. By invoking and subverting the *Bildungsroman* genre through his narration of this young slum-dweller's troubled coming of age, Abani demonstrates the paralysing imbrication of the local, national and global discourses of development that collide at the urban margins. While *GraceLand* suggestively critiques the instability and inconsistency of the postcolonial nation-state, Abani reveals the promise of cultural transnationalism to be equally circumscribed by Elvis' immersion in a global economic system that perpetuates his marginalisation.

Chapter 3 turns from Third World development to First World urban regeneration. However, these apparently disparate processes prove to be intimately connected. Narrated by Sepha, a profoundly alienated Ethiopian exile, Dinaw Mengestu's *The Beautiful Things That Heaven Bears* (2007) portrays the extensive redevelopment of the downtrodden Logan Circle neighbourhood in Washington, D.C. during the mid-1990s. I argue that by dramatising and enacting what Frederic Jameson (1988) terms 'global cognitive mapping', the novel demonstrates that gentrification is not only a strategic manoeuvre with decided local effects, but also a symptom of a globally widespread mode of order-building that rests on the violent elimination of wasted urban populations.

While the preceding chapters examine novels which seek to make urban waste visible, Chapter 4 studies an ostensibly non-fiction text which problematises this literary endeavour. Reading Suketu Mehta's *Maximum City* (2004) alongside Indian film and a literary tradition of migrant writing about Indian urban waste, this chapter argues that this kaleidoscopic text offers a persuasive critique of the contradictory dynamics of visibility that intersect the margins of contemporary Bombay. Mehta's distinctive ethnographic gaze reveals how the city's most devalued residents – criminals, slum-dwellers, sex workers – are caught within a disjunctive matrix of exposure and invisibility over which they have limited control. However, the inescapably gendered perspective Mehta brings to bear on Bombay's female bar dancers intensifies their vulnerability, troubling his representational empowerment of the urban margins.

Chapter 5 examines two artistic projects that subvert the paradoxical visibility of urban waste through their innovative 'remediations' of discarded things, degraded spaces and devalued people. Both a corrective and creative process, the conversion of urban waste into visual and performance art productively defamiliarises the uneven modes of waste production, disposal and management that underpin contemporary urban ecosystems. Looking at the recent documentary films *Waste Land* (2010) – an account of modern artist Vik Muniz' transformation of Rio de Janeiro trash pickers and refuse into mixed-media portraits – and *Trash Dance* (2013) – the story of American choreographer Allison Orr's staging of a dance performance using the garbage workers and trucks of Austin, Texas – this chapter considers the ways in which these self-reflexive visual texts acknowledge and avoid the risks inherent in aestheticising waste, instead demonstrating how creative representations of the urban margins offer powerful tools for social and environmental engagement.

Notes

1 Robert Neuwirth's *Shadow Cities* (2005) avoids the term altogether in its comparative analysis of 'squatter communities' across four continents. 'Slum is a loaded term', Neuwirth writes, 'and its horizon of emotion and judgment comes from outside. . . . To call a neighbourhood a slum establishes a set of values – a morality that people outside the slum share – and implies that inside those areas, people don't share the same principles. . . . It is a totalising word – and the whole, in this case, is the false' (pp. 16–17).

2 See Neville and Villeneuve (2002), Hawkins and Muecke (2003), Cohen and Johnson (2005), Hawkins (2006), Gee (2010), Sullivan (2012), Scanlan and Clark (2013). Relevant journal special issues include: *Waste and Abundance. SubStance* vol. 37, no. 2 (2008); *Waste. Iowa Journal of Cultural Studies* vols. 10/11 (2009); Special cluster on 'waste', *ISLE* vol. 20, no. 3 (2013). For current news, analyses and announcements pertinent to the field, *discardstudies.com* is a useful resource. As described on its 'About' page, the website provides a 'gathering place for scholars, activists, environmentalists, students, artists, planners, and others whose work touches on themes relevant to the study of waste and wasting'.

3 Drawing largely on examples from American literature and visual art, Patricia Yaeger's 'The Death of Nature and the Apotheosis of Trash' (2008) persuasively argues for 'the power of waste at the center of contemporary literature' (p. 331); see also her 2010 article, 'Sea Trash'. See also William G. Little, *The Waste Fix: Seizures of the Sacred from Upton Sinclair to The Sopranos* (2002), and Mary Foltz, *An Ethics of Waste: Twentieth-Century American Literature and Excremental Culture* (2009). Focusing on the Western canon, Susan Signe Morrison's *The Literature of Waste: Material Ecopoetics and Ethical Matter* (2015) suggests that literary waste has the potential to foster 'cultural harmony and understanding' (p. 9).

4 See also Ryan Connor's 'Regimes of Waste' (2013) for an analysis of twentieth-century African fiction that draws on Harrow's trash paradigm.

5 See Agozino (2013) for a review that strongly objects to what the author perceives as Harrow's reinforcement of degrading African stereotypes through his use of 'trash' as the organising principle for his study.

6 As Lincoln explains: 'Literature is thus, in its own way, a "trash heap": a site at which waste accumulates, but also (as in many real-world African landfills) a rich source of value, creativity, nutrition, and even surprising beauty, for those who have the skill to recognise them' (2008, pp. 4–5).

7 See Slaymaker (2001), Deloughrey *et al.* (2005), Deloughrey (2007), Marzec (2007), Caminero-Santangelo (2007), Cilano and Deloughrey (2007), Huggan and

Tiffin (2010), Mukherjee (2010), Roos and Hunt (2010), Wright (2010), Caminero-Santangelo and Myers (2011), Deloughrey and Handley (2011), Nixon (2011), Bartosch (2013), Caminero-Santangelo (2014) and Deloughrey, Didur and Carrigan (2015).

8 Notable exceptions include Elizabeth DeLoughrey (2010), who locates in postcolonial portrayals of the 'heavy waters of the Atlantic' an evocative site for 'exploring waste as a constitutive process and product of the violence of Atlantic modernity' (pp. 710–11).

9 See, for example, Alaimo (2010), Bennett (2010), Coole and Frost (2010), Iovino and Oppermann (2012).

10 Focusing on bodily wastes in particular, Susan Signe Morrison's *Excrement In the Late Middle Ages: Sacred Filth and Chaucer's Fecopoetics* (2008) relatedly calls attention to the 'cultural politics of excrement' articulated in Medieval English writing (p. 2). For further analysis of the social and sexual implications of scatology in English literature from the Middle Ages to the Early Modern period, see Peter J. Smith, *Between Two Stools: Scatology and its Representation in English Literature, Chaucer to Swift* (2012), and Will Stockton, *Playing Dirty: Sexuality and Waste in Early Modern Comedy* (2011).

11 For a thorough appraisal of Douglas' 'sociological conservatism', see Richard Fardon's *Mary Douglas: An Intellectual Biography* (1999).

12 For a related example of how the US government bolsters contemporary imperialism on the home front by discursively constructing the American West and its inhabitants as 'waste,' see John Beck's *Dirty Wars: Landscape, Power, and Waste in Western American Literature* (2009).

13 Here I'm drawing on Homi Bhabha's 'Of Mimicry and Man', in which he famously argues that the ambivalence of colonial discourse can be seen in its expressed 'desire for a reformed, recognisable Other, *as a subject of a difference that is almost the same, but not quite*' (1994, p. 86).

14 See, for example, Robert Home's history of colonial urbanism, *Of Planning and Planting* (2013) for a useful analysis of how discourses of sanitation motivated racially segregated town plans in Britain's tropical colonies.

15 For an instructive survey of the conflicts and commonalities between postcolonial and globalisation studies, see Krishnaswamy (2008).

16 See Friedmann (1986), Abu-Lughod (1999), Sassen (2001; 2012).

17 See Robinson (2006), King (2007), Nuttall and Mbembé (2008).

References

Abani, C. (2004) *GraceLand*. New York: Picador.

Abu-Lughod, J. (1999) *New York, Chicago, Los Angeles: America's Global Cities*. Minneapolis, MN: U of Minnesota P.

Agozino, B. (2013) 'Kenneth Harrow's *Trash*: Garbage In, Garbage Out', *Pambazuka*, 16 May 2013 [Online]. Available at http://www.pambazuka.net/en/category.php/books/87407 (Accessed 4 April 2014).

Alaimo, S. (2010) *Bodily Natures: Science, Environment, and the Material Self*. Bloomington, IN: Indiana UP.

Anderson, W. (1995) 'Excremental Colonialism: Public Health and the Poetics of Pollution', *Critical Inquiry*, vol. 21, no. 3: pp. 640–69.

Angotti, T. (2006) 'Apocalyptic Anti-Urbanism: Mike Davis and his Planet of Slums', *International Journal of Urban and Regional Research,* vol. 30, no.4: pp. 961–67.

Armah, A. K. (1968) *The Beautyful Ones Are Not Yet Born*. Oxford: Heinemann.

Bartosch, R. (2013) *Ecocriticism and the Event of Postcolonial Fiction*. Amsterdam: Rodopi.

Bauman, Z. (2004) *Wasted Lives: Modernity and its Outcasts*. Cambridge: Polity Press.

Beck, J. (2009) *Dirty Wars: Landscape, Power, and Waste in Western American Literature.* Lincoln, NE: U of Nebraska P.

Bennett, J. (2010) *Vibrant Matter: a Political Ecology of Things.* Durham, NC: Duke UP.

Beukes, L. (2010) *Zoo City.* Johannesburg: Jacana.

Bhabha, H. K. (1994) 'Of Mimicry and Man: The Ambivalence of Colonial Discourse', in *The Location of Culture.* London: Routledge: pp. 85–92.

Caminero-Santangelo, B. (2007) 'Different Shades of Green: Ecocriticism and African Literature', in Olaniyan, T. and Quayson, A. (eds.) *African Literature: An Anthology of Criticism and Theory.* Oxford: Blackwell: pp. 698–706.

Caminero-Santangelo, B. (2014) *Different Shades of Green: African Literature, Environmental Justice and Political Ecology.* Charlottesville, VA: U of Virginia P.

Caminero-Santangelo, B. and Myers, G. (eds.) (2011) *Environment at the Margins: Literary and Environmental Studies in Africa.* Athens, OH: Ohio UP.

Chamoiseau, P. (1997 [1992]) *Texaco* (trans. from French and Creole by R-M Réjouis and V. Vinokurov). New York: Vintage.

Cilano, C. and Deloughrey, E. (2007) 'Against Authenticity: Global Knowledges and Postcolonial Ecocriticism', *ISLE*, vol. 14, no. 1: pp. 71–87.

Çinar, A. and Bender, T. (eds.) (2007) *Urban Imaginaries: Locating the Modern City.* Minneapolis, MN: U of Minnesota P.

Cohen, W. A. and Johnson, R. (eds.) (2005) *Filth: Dirt, Disgust, and Modern Life.* Minneapolis, MN: U of Minnesota P.

Connor, R. (2013) 'Regimes of Waste: Aesthetics, Politics, and Waste from Kofi Awoonor and Ayi Kwei Armah to Chimamanda Adichie and Zeze Gamboa', *Research in African Literatures*, vol. 44, no. 4: pp. 51–68.

Coole, D. and Frost, S. (eds.) (2010) *New Materialisms: Ontology, Agency, and Politics.* Durham, NC: Duke UP.

Crosby, A. (1986) *Ecological Imperialism: The Biological Expansion of Europe, 900–1900.* Cambridge: Cambridge UP.

Cunningham, D. (2007) 'Slumming It: Mike Davis's Grand Narrative of Urban Revolution', *Radical Philosophy*, Issue no. 142 (Mar/Apr 2007): pp. 8–18.

Davis, M. (2005) *Planet of Slums.* London: Verso.

Deloughrey, E. (2007) *Routes and Roots: Navigating Caribbean and Pacific Island Literatures.* Honolulu: U of Hawai'i P.

Deloughrey, E. (2010) 'Heavy Waters: Waste and Atlantic Modernity', *PMLA*, vol. 125, no. 3: pp. 703–12.

Deloughrey, E., Didur, J. and Carrigan, A. (2015) *Global Ecologies and the Environmental Humanities: Postcolonial Approaches.* New York: Routledge.

DeLoughrey, E., Gosson, R. and Handley, G. (eds.) (2005), *Caribbean Literature and the Environment: Between Nature and Culture.* Charlottesville: U of Virginia P.

DeLoughrey, E. and Handley, G. (eds.) (2011) *Postcolonial Ecologies: Literatures of the Environment.* Oxford: Oxford UP.

Donald, J. (1999) *Imagining the Modern City.* Minneapolis, MN: U of Minnesota P.

Douglas, M. (2005 [1966]) *Purity and Danger: An Analysis of Concept of Pollution and Taboo.* London: Routledge.

Esty, J. D. (1999) 'Excremental Postcolonialism', *Contemporary Literature*, vol. 40, no. 1: pp. 22–59.

Esty, J. D. (2007) 'The Colonial Bildungsroman: *The Story of an African Farm* and the Ghost of Goethe', *Victorian Studies*, vol. 49, no. 3: pp. 407–30.

Fanon, F. (2004 [1961]) *The Wretched of the Earth* (trans. from French by R. Philcox). New York: Grove Press.

Fardon, R. (1999) *Mary Douglas: An Intellectual Biography*. London: Routledge.

Foltz, M. (2009) *An Ethics of Waste: Twentieth-Century American Literature and Excremental Culture*. PhD Thesis, State University of New York, Buffalo, NY.

Friedmann, J. (1986) 'The World City Hypothesis', *Development and Change*, vol. 17, no. 1: pp. 69–84.

Gamber, J. B. (2012) *Positive Pollutions and Cultural Toxins: Waste and Contamination in Contemporary U.S. Ethnic Literatures*. Lincoln, NE: U of Nebraska Press.

Gee, S. (2010) *Making Waste: Leftovers and the Eighteenth-century Imagination*. Princeton, NJ: Princeton UP.

Grove, R. (1995) *Green Imperialism: Colonial Expansion, Tropical Island Edens and the Origins of Environmentalism, 1600–1860*. Cambridge: Cambridge UP.

Harrow, K. (2013) *Trash: African Cinema from Below*. Bloomington, IN: Indiana UP.

Harvey, D. (1989) *The Condition of Postmodernity: An Enquiry into the Origins of Cultural Change*. Oxford: Blackwell.

Hawkins, G. (2006) *The Ethics of Waste: How We Relate to Rubbish*. Lanham, MD: Rowman & Littlefield.

Hawkins, G. and Muecke, S. (2003) *Culture and Waste: The Creation and Destruction of Value*. Oxford: Rowman & Littlefield.

Holston, J, and Appadurai, A. (1996) 'Cities and Citizenship', *Public Culture* vol. 8, no. 2: pp. 187–204.

Home, R. (2013) *Of Planning and Planting: The Making of British Colonial Cities*. 2nd edn. Abingdon: Routledge.

Huggan, G. and Tiffin, H. (2010) *Postcolonial Ecocriticism: Literature, Animals, Environment*. London: Routledge.

Iovino, S., and Oppermann, S. (2012) 'Theorizing Material Ecocriticism: A Diptych', *Interdisciplinary Studies in Literature and Environment*, vol. 19, no. 3: pp. 448–75.

Jameson, F. (1988) 'Cognitive Mapping', in Nelson, C and Grossberg, L (eds.) *Marxism and the Interpretation of Culture*. Chicago, IL: U of Illinois P: pp. 347–60.

Johnson, E. (2012) 'The Poetics of Waste: Medieval English Ecocriticism', *PMLA*, vol. 127, no. 3: pp. 460–76.

King, A. (2007) 'Boundaries, Networks, and Cities: Playing and Replaying Diasporas and Histories', in Çinar, A. and Bender, T. (eds.) *Urban Imaginaries: Locating the Modern City*. Minneapolis, MN: U of Minnesota P: pp. 1–16.

Krishnaswamy, R. (2008) 'Connections, Conflicts, Complicities', in Krishnaswamy, R. and Hawley, J. C. (eds.) *The Postcolonial and the Global*, Minneapolis, MN: U of Minnesota P: pp. 2–21.

Kristeva, J. (1982 [1980]) *Powers of Horror: An Essay on Abjection* (trans. from French by L. S. Roudiez). New York: Columbia UP.

Lefebvre, H. (1991 [1974]) *The Production of Space* (trans. from French by D. Nicholson-Smith). Oxford: Blackwell.

Lincoln, S. (2008) *Expensive Shit: Aesthetic Economies of Waste in Postcolonial Africa*. PhD Thesis, Duke University, NC.

Lioi, A. (2007) 'Of Swamp Dragons: Mud, Megalopolis, and a Future for Ecocriticism', in Merrill, A. *et al.* (eds.) *Coming into Contact: Explorations in Ecocritical Theory and Practice*. Athens, GA: U of Georgia P, pp. 17–38.

Lionnet, F. and Shi, S. (2005) *Minor Transnationalism*. Durham, NC: Duke UP.

Little, W. G. (2002) *The Waste Fix: Seizures of the Sacred from Upton Sinclair to The Sopranos*. New York: Routledge.

Martinotti, G. (1999) 'A City for Whom? Transients and Public Life in the Second-Generation Metropolis', in Beauregard, R. and Body-Gendrot, S. (eds.) *The Urban Moment: Cosmopolitan Essays on the Late 20th Century City*. London: Sage, pp. 155–183.

Marzec, R. P. (2007) *An Ecological and Postcolonial Study of Literature: From Daniel Defoe to Salman Rushdie*. New York: Palgrave Macmillan.

Mehta, S. (2004) *Maximum City: Bombay Lost and Found*. New York: Vintage.

Melosi, M. (2005) *Garbage in the Cities*. 2nd edn. Pittsburgh, PA: U of Pittsburgh P.

Mengestu, D. (2007) *The Beautiful Things That Heaven Bears*. New York: Riverhead.

Metcalf, T. R. (1995) *Ideologies of the Raj*. Cambridge: Cambridge UP.

Moore, S. A. (2012) 'Garbage Matters: Concepts in New Geographies of Waste', *Progress in Human Geography*, vol. 36, no. 6: pp. 780–99.

Morrison, S. S. (2008) *Excrement In the Late Middle Ages: Sacred Filth and Chaucer's Fecopoetics.* New York: Palgrave Macmillan.

Morrison, S. S. (2015) *The Literature of Waste: Material Ecopoetics and Ethical Matter*. New York: Palgrave Macmillan.

Mukherjee, U. P. (2010) *Postcolonial Environments: Nature, Culture and the Contemporary Indian Novel in English*. London: Palgrave Macmillan.

Naipaul, V. S. (1964) *An Area of Darkness*. London: Andre Deutsch.

Neuwirth, R. (2005) *Shadow Cities: A Billion Squatters, A New Urban World*. New York: Routledge.

Neville, B. and Villeneuve, J, (eds.) (2002) *Waste-site Stories: The Recycling of Memory*. Albany, NY: SUNY Press.

Nixon, R. (2005) 'Environmentalism and Postcolonialism', in Loomba, A. *et al*. (eds.) *Postcolonial Studies and Beyond*. Durham, NC: Duke UP, pp. 233–51.

Nixon, R. (2011) *Slow Violence and the Environmentalism of the Poor*. Cambridge, MA: Harvard UP.

Nuttall, S. and Mbembé, A. (2008) *Johannesburg: The Elusive Metropolis*. Durham, NC: Duke UP.

Pratt, M. L. (1992) *Imperial Eyes: Travel Writing and Transculturation*. London: Routledge.

Quartey, K. (2011) *Children of the Street*. New York: Random House.

Robinson, J. (2006) *Ordinary Cities: Between Modernity and Development*. London: Routledge.

Roos, B. and Hunt, A. (eds.) (2010) *Postcolonial Green: Environmental Politics and World Narratives*. Charlottesville, VA: U of Virginia P.

Said, E. W. (1978) *Orientalism*. New York: Random House.

Sassen, S. (2001) *The Global City: New York, London, Tokyo*. 2nd edn. Princeton, NJ: Princeton UP.

Sassen, S. (2012) *Cities in a World Economy*. 4th edn. Thousand Oaks, CA: Pine Forge Press.

Scanlan, J. (2005) *On Garbage*. London: Reaktion.

Scanlan, J. and Clark, J. (eds.) (2013) *Aesthetic Fatigue: Modernity and the Language of Waste.* Newcastle: Cambridge Scholars.

Seshadri-Crooks, K. (1995) 'At the Margins of Postcolonial Studies', *ARIEL*, vol. 26, no. 3: pp. 47–71.

Slaymaker, W. (2001) 'Ecoing the Other(s): The Call of Global Green and Black African Responses', *PMLA*, vol. 116, no. 1: pp. 129–44.

Smith, M. P. (2001) *Transnational Urbanism: Locating Globalization*. Oxford: Blackwell.

Smith, P. J. (2012) *Between Two Stools: Scatology and its Representation in English Literature, Chaucer to Swift*. Manchester: Manchester UP.

Stockton, W. (2011) *Playing Dirty: Sexuality and Waste in Early Modern Comedy*. Minneapolis. MN: U of Minnesota P.

Stoler, A. L. (2008) 'Imperial Debris: Reflections on Ruins and Ruination', *Cultural Anthropology*, vol. 23, no. 2: pp. 191–219.

Sullivan, H. (2012) 'Dirt Theory and Material Ecocriticism', *Interdisciplinary Studies in Literature and Environment*, vol. 19, no. 3: pp. 515–31.

Trash Dance (2013). Film. Directed by Andrew Garrison. [DVD]. USA: PBS International.

Waste Land (2010). Film. Directed by Lucy Walker. [DVD]. USA: Arthouse.

Wright, L. (2010) *Wilderness into Civilized Shapes: Reading the Postcolonial Environment*. Athens, GA: U of Georgia P.

Yaeger, P. (2008) 'The Death of Nature and the Apotheosis of Trash; Or, Rubbish Ecology', *PMLA*, vol. 123, no. 2: pp. 321–39.

Yaeger, P. (2010) 'Sea Trash, Dark Pools, and the Tragedy of the Commons', *PMLA*, vol. 125, no. 3: pp. 523–45.

1 'Anything could turn out to be something'

Gleaning slum history in Patrick Chamoiseau's *Texaco*

In early February 2009, the Caribbean screening of an hour-long special news report exposing the ongoing industrial monopoly of 'The Last Masters of Martinique' catalysed the spread of a general strike from Guadeloupe to Martinique ('Les Derniers Maîtres').[1] The programme highlighted the neocolonial economic control of Martinique's white Creole minority over the island's key industries, adding racial tensions to workers' existing calls for pay increases and reduced water and electricity bills. For over a month, gas stations, supermarkets and basic services were shut down on both islands. Martinique's capital, Fort-de-France, became the theatre on which the strikers' demands were played out with thousands marching through the city to voice their protests. The gradual accumulation of garbage in the streets caused particular offence to the hoteliers, travel agents and business owners who lost income due to cancelled tourist visits. These piles of stinking trash materialised the workers' refusal to be cast aside by an economic and social system that devalues their skills and productivity. A number of the strikers' demands were met with the signing of a resolution in March 2009, which granted salary increases to the lowest-paid workers.[2] However, rates of poverty and unemployment on the island remain high.

The strike points to Martinique's ambiguous postcolonial status. First colonised in 1635, the island has never claimed full independence from France. In 1946, Martinique was designated as a French 'overseas province' (*département d'outre-mer*), securing its ongoing formal connection to the French state. Although departmentalisation has arguably contributed to better standards of living in Martinique by comparison with neighbouring Caribbean islands such as Haiti, critics argue that this arrangement has furthered the island's cultural assimilation and economic dependence.[3]

Patrick Chamoiseau is one of a number of prominent Martinican activist-writers to challenge the island's neocolonial status quo, the tangible expression of which he locates in its rapid urbanisation.[4] In *Écrire en pays dominé*, an account of intellectual life in his 'dominated' home country, Chamoiseau laments the economic and environmental transformations that have been facilitated by French loans and investments, including: 'constructions in king-concrete, windows, electricity, traffic lights, television, car mania, triumphant low-income housing, sewage, Social Security, welfare, planes, roads and highways, schools, clothing stores, hotels, supermarkets, advertisements' (Chamoiseau 1997a, pp. 69–70, my translation). At first glance, this catalogue of developments seems to include many beneficial enhancements to the island's transport, residential and commercial infrastructures. However, it is in precisely this apparently innocuous functionality that Chamoiseau finds the 'furtive domination' of the island by France (Chamoiseau 1997a, p. 219). Determined to a large extent by European economic interests, the form and feel of Martinican urban life is closely bound to the metropole. Everyday urban mobilities, activities and interactions are shaped from afar by distant stakeholders. If, as Anthony King argues, 'how people build affects how people think', French sponsorship of Martinique's *bétonisation* ('cementing over') can be seen to extend France's colonial policy of assimilation into the present (King 1984, p. 99). Based on the coerced integration of the colonised into the French nation, the assimilation doctrine demands homogeneity at the expense of preserving cultural distinctions. As Robert Young puts it, this model of colonialism 'saw difference, and sought to make it the same' (Young 2001, p. 32). Martinique's recent urbanisation indicates the island's profound structural and cultural ties to France, which extend far beyond a formal administrative relationship between 'centre' and 'periphery'.

The rapid development of Martinique materialises the ongoing suppression of the island's painful past. Although the French government has recently made public efforts to acknowledge the legacy of French colonial slavery in the Caribbean, Martinican history continues to be subordinated to metropolitan historical discourse. As Reneé Gosson points out, the industrial 'cover-up' of the island compounds the cultural alienation of its population by disrupting access to the landscape's commemorative potential (Gosson 2006, p. 226). In *Caribbean Discourse*, Edouard Glissant emphasises the importance of this distinctive environmental archive to the creation of a unified Caribbean identity, explaining that 'our landscape is all monument: its meaning can only be traced on the underside. It is all history' (Glissant 1989, p. 11). As neocolonial industry, especially tourism, continues to estrange Martinicans from their physical surroundings, the historical record embedded in the landscape becomes even harder to retrieve.[5]

Chamoiseau's 1992 novel *Texaco* redresses this historical impasse through its elaboration of a marginal urban imaginary that elicits an alternative archive of the island's colonial past from the different forms of waste that proliferate in tandem with Martinique's apparent development.[6] Instead of retreating to an idealised or exaggerated rural environment, Chamoiseau consolidates Martinican identity by imagining the circumstances leading to the foundation of the eponymous slum Texaco, named after the oil company that owns the land on which the first of its shacks (or 'hutches') is built. In contrast to earlier social realist novels of Caribbean slum life, *Texaco* is a polyphonic and self-reflexive text that continually weaves together fictional and real-life events.[7] Chamoiseau himself appears in the novel as his authorial alter ego Oiseau de Cham, who is deeply sympathetic to the plight of the Texaco slum-dwellers. He explains in his afterword that he discovered the slum in the mid-1980s while conducting research for a previous novel. The main body of the text is his transcription of the oral history of the slum as told to him by its founder Marie-Sophie Laborieux, an elderly and formidable *femme-matador* who has seen off many violent and bureaucratic challenges to its existence.[8] She has already told this oral history once before to the unnamed 'Urban Planner' who was sent to survey Texaco before its planned demolition by the city authorities. By telling him the community's history she successfully defends the slum from being razed to the ground. Excerpts from the Urban Planner's notes are interspersed throughout the novel, in which he repeatedly expresses his admiration for the vitality and creativity of the slum-dwellers.

For those long stretches of Texaco's history preceding her own birth, Marie-Sophie draws on stories that her father Esternome passed onto her before his death. Born into slavery on one of Martinique's sugar plantations during the early nineteenth century, Esternome is freed when he saves the plantation owner from being killed by a maroon. He migrates to the city of Saint-Pierre, which is destroyed by the eruption of Mount Pelée in 1902. This forces him to move to Fort-de-France where he meets Marie-Sophie's mother, Idoménée, also a former slave. After her parents' deaths, Marie-Sophie survives poverty, displacement and rape while performing many gruelling jobs in the city. She eventually founds Texaco in 1950 and fights against its demolition until her death forty years later. At the close of the novel, the Fort-de-France authorities formally acknowledge the slum through an organised upgrading programme that incorporates it into the city's infrastructure.

Early commentators focused on the novel's inventive language and experimental form, none more exuberantly than fellow Caribbean writer Derek Walcott whose 'Letter to Chamoiseau' in the *New York Review of Books* celebrated its 'combined triumph of the Creole language and of French orthography', before going on to compare Chamoiseau's 'masterpiece' to Joyce's *Ulysses* (Walcott 1997).[9] Following its largely favourable critical reception on initial publication, *Texaco* was awarded the prestigious *Prix Goncourt* later that year.[10] Its translation into English in 1997 opened the novel up to a wider readership, both academic and mainstream. The subject of numerous critical articles and frequently included on college literature syllabi, *Texaco* has entered the postcolonial literary canon. In recent years, three monographs dealing with Chamoiseau's oeuvre have appeared, foregrounding his treatment of space, memory and form respectively.[11] This chapter extends all three emphases in its particular attention to Chamoiseau's ecological narrative method, which highlights the inextricability of human and environmental histories throughout the novel.

Marginalised within the city and peripheral to the official French history in which that of the entire island is subsumed, Texaco's struggle for inclusion in Fort-de-France bears metonymic relation to the unequal balance of discursive and economic power which binds Martinique to France. However, in his unique attempt to historicise slum existence, Chamoiseau depicts a local experience with increasing global resonance. While Fort-de-France is much smaller than many developing cities (the total population of the entire island is only just over 430,000), Chamoiseau's portrayal of Texaco provides what critic Michael Rubenstein calls an 'exemplary slum-urban vision' that vividly conveys the material hardships faced by today's 'new urban subject' (Rubenstein 2008, p. 35). Ashley Dawson concurs, describing the Texaco community's displacement as 'paradigmatic' in an essay that highlights the direct connection between the deprivation of the slum-dwellers and the inequity of global capitalism (Dawson 2004, p. 18). With the continuing expansion of cities around the world, their residents face heightened pressures: an increasing gap between rich and poor, unstable infrastructure, precarious housing, industrial and domestic pollution, scarce employment, state and criminal violence. In his attention to the problems faced by the island's internal diaspora – the displaced urban poor who struggle to attain not only shelter, but also a sense of belonging in contemporary Martinique – Chamoiseau depicts an increasingly common urban existence.

In order to survive, the Texaco slum-dwellers must learn to live in and from the urban waste to which they have been condemned by their historical legacy and the neocolonial economy. They do so by appropriating the rural practice of gleaning for their urban existence. Based on the inventive reuse of discarded or forgotten material, gleaning provides them with second-hand materials for shelter and sustenance. Although some critics have noted an equation between the countryside and cultural authenticity in Chamoiseau's early works, the slum-dwellers' urban revival of this agrarian foraging practice diminishes a misleading city–country binary which falsely holds these two spheres in opposition.

Gleaning is not only a practical everyday necessity for the impoverished Texaco residents. The intertwined nature of Martinique's rural and urban histories is further demonstrated by the slum-dwellers' use of gleaning techniques in order to construct a legitimating historical account of Texaco's evolution. As founder of Texaco, Chamoiseau's principal narrator Marie-Sophie Laborieux is committed to defending the slum. Her diligent memory-gleaning resists the discursive disposal of its marginal urban history, serving as an effective mode of territorial reclamation.

Chamoiseau does not only thematise the slum-dwellers' gleaning, but he does so using a composite narrative form that depends on the same revaluation of scraps, leftovers and remnants that is essential to the slum-dwellers' survival. Often identified as an authorial *bricoleur*, Chamoiseau can be better understood as a *gleaner* of diverse narrative perspectives and styles. By revaluing that which has been discarded, rejected and overlooked, his formal appropriation of the slum-dweller's foraging techniques demonstrates an inclusive form of cultural invention and identification. However, the enduring tension between Chamoiseau's stylistic innovation and the marginal condition that compels it prevents the amnesiac recuperation of the slum. The ironic gap between the novel's formal excess and thematic lack serves instead as an insistent reminder of the many losses and deprivations that precede Texaco's eventual incorporation into Fort-de-France.

Urban gleaning

In crafting his detailed narrative of the Texaco slum's emergence, Chamoiseau spotlights a frequently overlooked counterpart to the growth of Martinique's cities and businesses. The wasted spaces, objects and people that are produced by industrialisation trouble the notion that urban and commercial expansion indicates linear progress for the entire island. Not everyone or everything benefits equally, if at all, from such shifts. The slum-dwellers' reliance on gleaning for food and housing dramatises the inherent inequality of Martinique's putative development, the origins of which Marie-Sophie traces back to the exploitative plantation culture imposed by early French colonisers in the seventeenth century.

Chamoiseau evokes the environmental and ethical complacency of neocolonial industry through Marie-Sophie's description of the site on which the slum is gradually built. Early in the novel, she points to:

> A fenced space where a smell of stale oil permeated the soul. Texaco, the oil company which used to occupy that space and which had given its name to it, had left aeons ago. It had picked up its barrels, carted off its reservoirs, taken apart its tankers' sucking pipes, and left. Its tank trucks sometimes parked there, to keep one foot on the dear property. Around that abandoned space are our hutches, our very own Texaco, a company in the business of survival.
>
> (Chamoiseau 1997b, p. 24)

Polluted and unused, the defunct depot bears the imprint of Martinique's unwelcome dependence on foreign sources of energy and income. This was never a site of oil extraction, but rather importation and storage – like most Caribbean islands, Martinique has no oil reserves of its own. Industrialisation heightens the demand for fuel, but the small island market is evidently not enough to sustain long-term investment from the Texaco corporation. Contaminated by the literal trace of global business, this small space holds minimal appeal for anyone other than those forced into the role of trespassers with nowhere else to build their homes.

In naming the slum after the oil company, Marie-Sophie pays ironic tribute to the uneven global economy that has contributed to both her own poverty and that of her physical surroundings. Although the departed multinational was hardly a good custodian of the environment, Marie-Sophie refuses to idealise the slum-dwellers' relationship to the same land. As increasing numbers of residents start to build their 'hutches' on the abandoned oil reservoir, their immediate practical concerns override environmental sensitivity. Marie-Sophie notes that Texaco's new residents are themselves responsible for the production of new forms of urban waste:

> Everyone cleaned his hutch and around his hutch, leaving the rest of the laundry to time's washing. Everyone thought that, just like in the countryside, nature would digest the refuse. I had to tell them again and again that around City nature lost some of its strength and watched the garbage pile up along with us. But we had many other worries besides that question of garbage (the waves tossed it about, the mangrove swamp stiffened it into sinister scarecrows). I would have liked to put together a few hands to take care of all that, but there were a thousand wars to wage merely to exist. After that, we learned, between the flies and mosquitoes, the smells and the miasmas, about living as straight-backed as possible.
>
> (Chamoiseau 1997b, p. 320)

According to Oiseau de Cham's authorial afterword, the concept of 'City' here and throughout the novel 'designates not a clearly defined urban geography, but essentially a content and therefore a kind of enterprise' (Chamoiseau 1997b, p. 386). City (translated from the Creole *l'En-ville*) is both a specific place – Fort-de-France 'proper' – but also an environment encompassing historical tensions, social currents and cultural attitudes. Here on the margins of City, the Texaco residents must live in close proximity to the rubbish that the more privileged residents of Fort-de-France freely put out of sight and mind. Despite their precarious position in the urban hierarchy, Marie-Sophie cautions against the slum-dwellers' unquestioning faith in the restorative power of an idealised 'nature'. She recognises the specificities of both the urban ecology and the rural environment, but she also notes the many natural processes that impact and enable City life: the waves, the mangroves and the buzzing insects. Contrary to the possessive, exploitative ideology of those who trade in natural resources, Marie-Sophie is alert to the inevitable contingency of human and environmental survival.

In order to eke out an existence in Texaco, Marie-Sophie and her neighbours must fully engage with the material and metaphorical waste that surrounds them. Unable to efficiently dispose of their garbage nor suppress the traumatic memories of their past, Marie-Sophie models an ethics of waste which helps them to survive. Confined to the margins of the city, their first priority is establishing shelter. As Marie-Sophie explains, 'in City, to be is first and foremost to possess a roof' (Chamoiseau 1997b, p. 275). Her decision to start building on the disused Texaco oil depot is challenged by its *béké* (white Creole) landowner with increasing violence and eventual assistance from the phonetically labelled 'seyaress' – the Compagnie Républicaine de Sécurité, a national police force rumoured to consist of 'Hitler's old henchmen whom the békés had ordered into the Colonies especially' (Chamoiseau 1997b, p. 306). Each time they destroy her hutch, Marie-Sophie insists on building and rebuilding it, defying their attempts to displace and ultimately erase her.

As she wanders the streets of Fort-de-France scouring the ground for useful building materials, Marie-Sophie demonstrates an empathetic attentiveness to rejected objects and imperfect products:

> I walked in the streets looking at the ground. From now on anything could be useful to me, a piece of string, the grace of a nail, an abandoned crate . . . anything could turn out to be something. My cunning bustling allowed me in the space of a few weeks to gather three crates, two new tin sheets, five slabs of cracked asbestos that a milato by the sea let me have on credit.
>
> (Chamoiseau 1997b, p. 299)

Marie-Sophie's 'cunning bustling' is a mode of urban survival that recalls the creative salvaging examined by Joanna Grabski in a recent study of Senegalese *récupération*, a mode of 'expressive production relying on materials culled from the urban environment' that gained particular prominence in Dakar during the 1990s (Grabski 2009, p. 8). As Grabski explains, such art demands a particular imaginative *vision*, a mode of perceiving the cityscape anew:

Looking for and gathering materials, the first step in récupération, relies on the conjunction of vision and mobility . . . [L]ooking is not a passive act of taking in or gaining inspiration; instead, it facilitates a subsequent act – that of imagination or, more specifically, *imagining objects otherwise.*

(Grabski 2009, pp. 12–13, my italics)

As both means of survival and mode of artistic creation, urban salvaging rests on the creative revaluation of the city's waste. By recognising the potential of her ostensibly degraded surroundings, noting that 'anything could turn out to be something', Marie-Sophie empathetically revisions that which has been cast aside in Fort-de-France (Chamoiseau 1997b, p. 299).

While Marie-Sophie's reuse of discarded materials recalls urban art practices, it also resonates with older agrarian foraging traditions, which were developed under conditions of deprivation and hunger. Stooped over, her eye trained on the ground, she assumes the classic downward-looking posture of a gleaner. Derived from the archaic French verb *glaner*, *glanage* or 'gleaning' describes the ancient agricultural practice of gathering leftover produce from newly harvested fields. A right traditionally accorded to the poor and landless, gleaning is referenced in many literary and cultural texts, from the Bible to Keats.[12] While an important source of income for labouring families in seventeenth- and eighteenth-century Europe, the practice declined with the advent of industrialisation, which led to more efficient harvesting techniques and the increasing privatisation of farming.[13] Displayed in the Paris salon as French gleaning rights were on the wane, French artist Jean-François Millet's 1857 painting *The Gleaners* offers an iconic European depiction of this practice. Critics have admired his realist portrayal of the 'quiet heroism' of three peasant women who quietly forage for food under the surveillance of a landowner on horseback in a manner not dissimilar to Marie-Sophie's urban salvaging (Crummy 1999, p. 18).[14]

As a mode of survival, postcolonial urban salvaging, is, however, a far more hazardous undertaking than that evoked by Millet's painting, which suggests a certain agrarian calmness and order. Despite the continuities of poverty, displacement and marginalisation that link nineteenth-century gleaning to the present, today's urban gleaners handle waste materials whose quantity and toxicity eclipses anything that Millet could have imagined. The 'five slabs of cracked asbestos', for example, that Marie-Sophie is grateful to salvage, harbour unseen toxins that she will inhale on a daily basis. Further examples of contemporary gleaning can be found in Accra's Agbogbloshie slum, which provides the dystopian setting for Kwei Quartey's *Children of the Street* (2011) as discussed in the previous chapter. The subject of a number of recent photographic series, the slum has developed around a dumping ground for hazardous e-waste, imported, often illegally, from richer Western countries. Residents here make their living not by scavenging food, but by salvaging and burning electronic devices to retrieve recyclable metals that they can then sell on. They are exposed to dangerously high levels of lead and mercury in the course of doing so. An untitled image from South African photographer Pieter Hugo's prize-winning *Permanent Error*

Figure 1.1 Hugo, Pieter (2010) *Untitled, Agbogbloshie Market, Accra, Ghana* from *Permanent Error*

(2011) evokes the pollution and peril in which the slum's residents live and work (see Figure 1.1).[15] While they share the downward gaze of Millet's gleaners, they are working independently of one another. The aprons in which Millet's peasants stored their gleanings have been replaced here by plastic bags – icons of the enduring waste of contemporary global consumer culture. Whereas Millet's gleaners operated under surveillance, these young men are alone, inextricable from the global economy yet lacking any formal protection or personal security.

Examining the history of gleaning foregrounds the social conditions which lead to its present-day necessity. Reflecting on today's apparent acceptance of widespread need, Donna Haraway (2009) notes that 'gleaning is tied to hunger. . . Hunger is tied to poverty and poverty is not a natural disaster but a political arrangement'. Understanding Marie-Sophie as a gleaner prevents the naturalisation of the specific social and historical circumstances that have led to the dispossession of the Texaco residents. Marie-Sophie's experience is not an exclusively modern urban phenomenon, but a socially marginalised existence with a precedent in plantation society. Her gleaning suggests a continuity between the deprivations of her enslaved grandparents and her own struggles in Fort-de-France. Chamoiseau's depiction of Marie-Sophie as a contemporary urban gleaner thus offers a reminder of the long history of displacement that precedes the struggles of the Texaco community.

Some critics have argued that *Texaco* idealises this history in its sympathetic and often humorous depiction of the slum's genealogy. In a largely sceptical analysis of 'Chamoiseau's radiant writing of the Creole town', Mary Gallagher challenges the 'spatial creolisation' which characterises 'urban space/time' by 'reproduc[ing] the structures of rural space/time' in the city (Gallagher 2002, p. 198, p. 189). For Gallagher, Chamoiseau's portrayal of the 'continuity between the plantation culture of the past and the urban future' problematically suggests the repetition of past rural struggles in the new space of the city (Gallagher 2002, p. 202). She perceives a nostalgia for an earlier 'oppositional moment' in Chamoiseau's depiction of the polarised, conflictual relationship between the city and the slum (Gallagher 2002, p. 197). According to Gallagher, the slum-dwellers' subversive survival strategies, such as gleaning, are an idealistic reproduction of the 'culture of ruse and resistance' that emerged on the plantations (Gallagher 2002, p. 197).

Although Gallagher detects an 'agenda of dichotomy' in Chamoiseau's writing, his ironic juxtaposition of the Urban Planner's increasing enthusiasm for the slum alongside Marie-Sophie's decline into old age and alcoholism undermines the naïve celebration of its marginal status Gallagher (2002, p. 198). In his notes, the Urban Planner effusively praises the oppositional vigour of the slum-dwellers, noting that 'Texaco remembers the play of forces between the hutches and the Big Hutch, between the plantation and the market town, the rural market town and the city' (Chamoiseau 1997b, p. 313). Yet whereas he excitedly proclaims the need to 'arouse a *countercity in the city*', expressing precisely the oppositional mentality to which Gallagher objects, the slum-dwellers themselves actually strive for incorporation (Chamoiseau 1997b, p. 361). Marie-Sophie initially assumes an unequivocally confrontational stance towards the city, asserting her intention to 'fight against City with a warrior's rage' (Chamoiseau 1997b, p. 271). Yet despite their many battles against the city, the Texaco residents express a persistent desire for recognition. They lament their neglect by the city's inhabitants and its authorities:

> We shoved our way about next to City, holding on to it by its thousand survival cracks. But City ignored us. Its activity, glances, the facets of its life (from every day's morning to the beautiful night neon) ignored us. We had vied for its promises, its destiny, we were denied its promises, its destiny. Nothing was given, everything was to be wrung out. . . . We saw City from above, but in reality we lived at the bottom of its indifference which was often hostile.
>
> (Chamoiseau 1997b, pp. 316–17)

If what Gallagher critiques in *Texaco* is a false or shallow spatial creolisation which insists on the conflict between rather than the productive intermixture of city, slum and countryside, the slum-dwellers' desire for incorporation arguably shares her preference for a more holistic urban space.

By patiently constructing shacks from gleaned building materials, Marie-Sophie and her fellow slum-dwellers gradually establish a proprietorial claim to the disused Texaco depot. As a small-scale means of rejecting local marginalisation, their gleaning provides a metonym for the larger struggle of Martinique to gain a foothold within a globalised economic system in which they are subject to metropolitan regulations and global inequities. However, the triumph of Texaco's eventual incorporation into the city proper is undercut by the Texaco *béké*'s unwelcome farewell visit to Marie-Sophie in which he reminds his longstanding adversary 'that the war was much larger and on that level he was not losing and never would' (Chamoiseau 1997b, p. 364). Although like many of the island's white landowners he is retreating from Fort-de-France to a more exclusive area, the *béké* retains the power of wealth even as his social and political visibility diminishes. Chamoiseau is clearly invested in the particular fate of Texaco, but here he also offers a pointed reminder of the ongoing global conflict within which the slum-dwellers are imbricated.

Metropolitan gleaning

In addition to its long history, gleaning is a survival practice with significant transnational resonance, as revealed through a contrapuntal analysis of metropolitan gleaning alongside *Texaco*'s Caribbean depiction. Attention to urban gleaning in Chamoiseau's novel thus reveals the complex and widely distributed responsibility for poverty in contemporary Fort-de-France, which must be simultaneously understood as a particular local instantiation of a much larger system of waste production. The broader context for the enduring dispossession of the Texaco slum-dwellers is readily illuminated by reading Agnès Varda's documentary *Les Glaneurs Et La Glaneuse* (*'The Gleaners and I'*) alongside Chamoiseau's novel. In this acclaimed 2001 film, Varda suggests that gleaning also offers a means of opposition for those situated within the metropole. Like Chamoiseau, her attention to urban waste practices disturbs the precarious distinction between margins and centre, suggesting instead their continuous imbrication.

Described in her own words as a 'wandering-road documentary', Varda's film traces French gleaning practices from their rural origins to their contemporary urban revival. Filmed entirely on hand-held camera, it features a voice-over and several appearances by the director herself, but Varda's primary interest is in the various gleaners who she interviews throughout France, some of whom gather leftovers from necessity, others for pleasure, artistic purposes or political protest. What unites this cast of migrants, artists and activists is a shared eccentricity. Whether imposed or intended, their social marginalisation offers a compelling new perspective on the forms of urban waste that globalisation generates within the First World. Extending Chamoiseau's critique of the hierarchical relationship between European 'centre' and Caribbean 'periphery', Varda explicitly documents the potential of gleaning to open up new circuits of environmental engagement and cultural exchange that exceed, extend and eschew those sanctioned by the global economy.

Recalling the origins of Texaco, Varda encounters a man who takes up gleaning in response to the wilful wastefulness of the multinational oil industry. He claims to have lived entirely on food salvaged from rubbish bins for over a decade. For this man, gleaning is a mode of political activism. He explains that he chooses to eat leftovers because he is frustrated by the environmental damage caused by consumerist capitalism. He finds in the Erika oil spill a charged example of what he acts against. Whereas the Texaco residents have little choice but to live with the effects of such environmental carelessness on a daily basis, Varda's subject is a horrified spectator who is particularly saddened by the death of many birds as a result of this oil spill. Gesturing to his neighbours, he tells Varda that 'all the others can die in their apartments under their rubbish. That's not a problem. Above all [I care about] the birds' (my translation). The misanthropic moral superiority that energises his gleaning clearly limits his political efficacy. For this man, gleaning facilitates a self-imposed social isolation, contrary to the community-building efforts of Marie-Sophie and her neighbours. However, his surprisingly passionate critique of the collateral damage caused to the environment by the oil industry underlines the global context in which local gleaning practices take place.

Towards the end of the film, Varda uncovers a more enduring model of transnational solidarity that gleaning facilitates. She identifies meeting Alain in a Parisian marketplace as one of her favourite encounters. She first notices him eating his gleanings on the spot as vendors pack up their stalls around him. Holding a master's degree in biology, as Varda later discovers, he seeks out a balanced, nutritional diet despite his meagre provisions. He explains that he lives in a men's shelter in an outlying suburb, but he does not say whether this is through choice or necessity. As their friendship develops after subsequent encounters at the same market, Varda films his daily commute to the city centre where he sells newspapers outside Montparnasse Station. Varda's admiration for Alain's humble lifestyle is clear from the respectful distance at which she films him. Compared to her other interviewees, she spends less time in close-up conversation with him, preferring to document what she terms his 'gestes modestes' ('modest gestures') at a remove.

Varda is particularly impressed by the nightly language and literacy classes Alain voluntarily teaches to his fellow shelter residents, many of whom are recent immigrants to France from Mali and Senegal. She captures a convivial classroom atmosphere in which Alain's students are learning to pronounce, amongst others, the French word for 'success'. Their shared laughter is undercut by a poignant sense that their shaky articulation of this concept marks its distance from their grasp – they try to translate it, but fail to come up with an equivalent. With a white Frenchman instructing African students in his language, this episode is initially reminiscent of a colonial classroom. However, Alain's instruction is an act of welcoming rather than enforced assimilation (the students attend class whenever they choose; they can come and go as they please). As the next chapter will discuss further, urban immigrants are frequently characterised as social 'waste' by hostile authorities, unwelcoming neighbourhoods and unwilling employers. In his effort to assist these men in their navigation of the French metropole, Alain extends his gleaning practices to redress their social alienation. Alain's evident humility and his shared social marginalisation ensure that this is an empathetic exchange, not a process of indoctrination or induction into perpetual social inferiority.

Varda's documentation of Alain's interaction with the immigrants exposes an often obscured dimension of metropolitan life. In keeping with recent efforts to theorise the discrepant globalities of such metropolitan centres as Paris from 'below', she reveals a productive form of what Françoise Lionnet and Shu-mei Shih (2005) term 'minor transnationalism'; in other words, an urban transcultural alliance which circumvents typically sanctioned interactions between first-world citizen and third-world immigrant. Putting Chamoiseau's gleaning narrative into dialogue with Varda's film further extends the transnational networks documented in the latter. In addition to creating local links between the Martinican countryside and the island's growing capital city, Marie-Sophie's gleaning suggests an affinity between Caribbean urban gleaners and French rural and metropolitan *glaneurs*. Read together, Chamoiseau and Varda usefully highlight a transatlantic waste practice that supplements the official connection between *département* and *metropole*. Although First and Third World cities tend to be addressed either separately or hierarchically within urban studies, a comparative reading of the gleaning practices which intersect both puts narratives of developed and developing cities into direct conversation, pointing the way to a comparative global history of urban poverty. A mode of micro-opposition with global reach, gleaning unlocks an alternative archive which reconfigures the relationship between periphery and metropole.

Memory gleaning

Aleida Assman notes that the purview of historians and gleaners does not typically coincide: 'Archives are repositories for things that are deemed worth preserving. As such, they have a reverse affinity with rubbish dumps, where things past are accumulated and left to decay' (Assman 2002, p. 71). However, like the visual artists whose *récuperation* work Assman examines, Chamoiseau revalues that which has been thrown away in order to craft an alternative archive, which reconfigures the well-documented colonial history of Martinique's relation to France and the rest of the world.

Unlike purposeful historical records, this history initially emerges by chance. Marie-Sophie retrieves items which have been discarded in the city dump in order to furnish her home; amongst them, 'a shimmering black casket decorated with a dragon which must have been one of those curiosities that some old person had brought here from Indochina . . .' (Chamoiseau 1997b, p. 280). In a useful analysis of this material object, Cilas Kemedijo suggests that this 'treasure' exposes 'the other side of globalisation' (Kemedijo 2001, p. 147, my translation). Discarded by an old Martinican soldier who had served for France during its colonial campaigns, this piece of furniture records the injustices of a colonial system that coerced colonised Caribbean subjects to fight against their Asian counterparts on behalf of a European power. Through such echoes, the dump becomes, as Assman further suggests, 'the emblem of a subversive counter-memory that cannot be controlled by the institutions of political power, figuring as a perpetual resource of creative energy' (Assman 2002, p. 81). Indeed, Qualidor, the watchman of the dump highlights its power in a monologue deployed to impress a potential lover. Not only does Qualidor show Péloponèse the many material treasures the dump has to offer, including 'dressers, windows, royal chairs, spoons made of heavy silver, thick old books, ropes, strings, pieces of hard plastic, disheveled rugs…' (Chamoiseau 1997b, p. 282), he also points out how they register the suppressed psychic trauma of the island:

> [T]here, look at…the fifty thousand longings of old blackmen whose bones look like tools, and over there…twelve syrian nightmares where bullets still fart, and there the pierced heart of a small syrian girl they wanted to marry off to a faraway syrian, and there Carib sea-tongues which surface from the ocean without saying why and which howl ugly screams into the conches' pink depth…and there…the misery I tell you of the coolie souls who have not found the boat home…
>
> (Chamoiseau 1997b, p. 281)

Here, that which has been discarded by privileged City residents is revealed to be more than a complacently overlooked material resource. The items which have been thrown away are expressive relics, symbolising Martinique's sidelined histories, which extend beyond the island to other peoples and places impacted by colonialism and its continuing effects.

Following Qualidor's insight, Marie-Sophie turns her gleaning skills to the recuperation of a historical narrative that can secure Texaco's long-term survival by asserting the slum-dwellers' long and legitimate claim to a place in the city. Traumatised by their colonial upbringing, Marie-Sophie's parents are both reluctant to tell her about the past:

> Mama, to avoid my questions, pretended she was wrestling with my hair she was braiding. She would drag the comb back like a farmer ploughing rocky ground who, surely you understand, does not have time to chitchat. Papa was more evasive. At my questions he slipped away smoother than a cool September wind. He would suddenly remember yams to pull out of the little puddles he had everywhere.
>
> (Chamoiseau 1997b, p. 34)

On the one hand figuratively, on the other literally, both Marie-Sophie's parents are figured as preoccupied landowners, trying to protect their crops. Marie-Sophie is cast as a needy peasant, who doggedly gathers the scraps of information which they let slip. Despite their extreme reticence, Marie-Sophie eventually collects enough 'bits of memory' from 'an inch of recollection here, a quarter of a word there' in order to tell the history of her own family and that of Texaco (Chamoiseau 1997b, p. 34). In his afterword, Oiseau de Cham admires Marie-Sophie's persistent memory-gleaning, noting that 'she had all her life run after her father's word and the rare words of Papa Totone and the morsels of our stories which the wind was sweeping away across the land just like that' (Chamoiseau

1997b, p. 387). Her gradual recuperation of her parents' memories painstakingly compiles a historical narrative that asserts the inherent value of a marginal urban community designated as waste.

As, bird-like, Marie-Sophie gathers up her parents' memories, she recalls her enslaved grandfather's obsession with blackbirds. By beginning her narrative with the story of his death, Marie-Sophie transforms the very inaccessibility of her ancestral history into its point of origin, making of the partiality of the historical record a generative example. A survivor of the Middle Passage, she learns that her grandfather was 'one of those men from Guinea' who performed his designated tasks as instructed without ever speaking (Chamoiseau 1997b, p. 36). All that breaks his silence is the utterance of an 'inaudible Low Mass' which he repeats while 'point[ing] to blackbirds and other blackbirds, blackbirds and other blackbirds' (Chamoiseau 1997b, p. 37). Marie-Sophie reveals the meaning of her grandfather's strange mantra to be a symbolic question: 'Until the end of his life the man had wondered how birds could be and how they could fly' (Chamoiseau 1997b, p. 38). If her grandfather perceived birds as an emblem of freedom, Marie-Sophie's gleaning highlights the tenuousness of her own liberty in postcolonial Martinique.

In relating the traumatic memory of her ancestor's literal disposal, Marie-Sophie exhibits a restorative care for the human refuse left in the wake of slavery. Chamoiseau, in turn, demonstrates attentiveness to marginalised histories by constructing the story of Texaco from the remnants of Marie-Sophie's ancestry. Suspected of poisoning livestock belonging to the plantation overseer, Marie-Sophie's grandfather is progressively deconstructed. First, physical torture reduces him to mere body parts:

> [T]he Béké got one of the most ferocious torturers of blackmen from the city to come and unleash all the resourcefulness of pliers on [her grandfather], to braise his blood, peel his skin, shatter his nails and some very sensitive bones, only to leave vanquished by this human wreck more mute than the dungeon itself.
>
> (Chamoiseau 1997b, p. 37)

He is then placed in an underground dungeon and left to rot, as Marie-Sophie explains:

> My papa knew the man from the dungeon to be his papa the day they pulled his remains covered with whitish fungus out of a foul-smelling hole. The Béké had it put on a pile of wood which he set on fire himself.
>
> (Chamoiseau 1997b, p. 38)

Marie-Sophie's grandfather is dehumanised, objectified and finally reduced to nothing through his uncaring cremation. Infuriated by his slave's inaccessible knowledge, the threatened Béké seeks complete mastery of his unruly body. Unable to control Marie-Sophie's grandfather, he attempts total elimination through violence.

Marie-Sophie's careful preservation of her grandfather's memory extends the tradition of slave resistance in which he actively participated. His knowledge of natural poisons and remedies, his vivid recollections of Africa and his apparent ability to be in two places at once mark him as a Mentoh, the embodiment of 'slave Power' (Chamoiseau 1997b, p. 52). Mistakenly described as 'necromancers, conjurers, sorcerers' by those inimical to their mysterious skills (Chamoiseau 1997b, p. 51), these 'men of strength' sought to 'preserve the remnants of humanity' amid the cruelties of slavery (Chamoiseau 1997b, p. 64). Marie-Sophie shares their impulse to salvage and protect her ancestral remains. She asserts their fundamental role in the history of Texaco, explaining that it was a Mentoh who persuaded her father Esternome to abandon the plantation after 'witnessing the burning of the remains, those of his own papa' (Chamoiseau 1997b, p. 50). In *Éloge de la Créolité*, Chamoiseau and his fellow créolistes lament the scarcity of testimonies to the 'obstinate progress of ourselves', during which 'those who stood up against the hell of slavery . . . left the fields for the towns, and spread among the colonial community to the point of giving it its strength in all respects, and giving it what we are today' (Bernabé, Chamoiseau and Confiant 1993, p. 98). From its origins in the resistance of her grandfather and his fellow Mentohs, Marie-Sophie's history of Texaco bears witness to the creation of an urban community forged in defiance of oppression and displacement.

The social vulnerability of Martinique's slaves and their ancestors is compounded by their exposure to a volatile natural environment. Marie-Sophie alerts the reader to the inextricability of the island's social and environmental histories when she notes the atmospheric changes that strangely coincide with her grandfather's suffering:

> During that man's endungeoning the plantation awoke under the gray down of a bird flapping its distraught wings in the mushy air of its sky. But no one cared. The down settled everywhere, enhancing the landscape with the color of a full moon. It must also have covered the lungs (for everyone sneezed) and lined their dreams with feathers (for some dreamt of yellow beaked humans flying in hurricanes).
>
> (Chamoiseau 1997b, p. 50)

The ominous reality behind this metaphor becomes clear some years later when, in 1902, the volcano of Mont-Pelée erupts, reducing Martinique's first capital city, Saint-Pierre, to rubble. A major event in the island's official national history, Chamoiseau filters its narration through the eyes of Esternome, whose first-hand account Marie-Sophie retells. The significance of the eruption is undermined by Esternome's initially careless response. Preoccupied with securing the affections of his lover Ninon, he ignores the volcanic warning signs: '[F]rom time to time the horizon became roaring and ... ash from the mountain suddenly floured the land, more and more often, for a longer and longer time' (Chamoiseau 1997b, p. 147). Reminiscent of the 'gray down' seen earlier, this description invites an association between the man-made and environmental crises that impact the island. When

Esternome enters the destroyed city following the eruption, he discovers a scene of biblical devastation that resonates with his father's death:

> A tide of ash. A deposit of still heat. The stone's red glow. Intact beings stuck to wall corners, going up in strings of smoke. Some were shriveled up like dried grass dolls. Children were savagely interrupted. Bodies undone, bones too clean, and oh how many eyes without looks.
>
> (Chamoiseau 1997b, p. 150)

Evoking the burning remains of Marie-Sophie's grandfather, the incinerated bodies of Saint-Pierre's inhabitants stand for the comprehensive erasure of the city's history. Esternome moves through the ruins in birdlike fashion, 'flapp[ing] his wings in silver ash' and taking 'side steps like a blackbird moving through glue' (Chamoiseau 1997b, p. 148, p. 149). He rescues two surviving prisoners who are trapped inside their cell, symbolically enacting the liberation which his own father lacked. In recounting this incident, Chamoiseau alludes to Martinican urban legend which claims that there were only two survivors of the earthquake. His revision of this myth suggests the malleability of historical narration. Chamoiseau bridges the historical void left by the death of Marie-Sophie's grandfather by layering competing historical accounts within a single narrative.

With Mont-Pelée's eruption, environmental violence overlays colonial violence, depriving Martinique's marginal urban dwellers of a grounding historical narrative. Chamoiseau dramatises their resulting disorientation though his depiction of Esternome's traumatised response to the devastating sight of the city laid to waste by the volcano. In the immediate aftermath of the eruption, Esternome suffers the troubling hallucinations characteristic of trauma victims. As he obsessively searches through the rubble of Saint-Pierre for his lost lover Ninon, he begins to see her everywhere: 'in each blown-up chest, in each puddle of flesh, on each pyre' (Chamoiseau 1997b, p. 150). The devastated city becomes the scene of psychic disintegration. In later life, Esternome is reticent about the horrors that he witnessed as a young man. When Marie-Sophie asks him about the volcanic disaster, she finds that her father 'covered it with the same stubborn silence he had kept his whole life concerning the old days in chains' (Chamoiseau 1997b, p. 149). His refusal to narrate is both traumatic symptom and mode of resistance. On the one hand, as Cathy Caruth has usefully observed, 'the traumatised carry an impossible history within them, or they become themselves the symptom of a history that they cannot entirely possess' (Caruth 1995, p. 5). Esternome, as silent witness, embodies the incomprehensible pain of the multi-layered, inextricable man-made and natural traumas of the Caribbean past. As Caruth argues, survivors' silence amounts to more than passive resistance. The choice to remain silent marks an active unwillingness to translate a traumatic event into a manageable, transferable narrative, which would diminish its 'essential incomprehensibility, the force of its *affront to understanding*' (Caruth 1995, p. 154). Esternome explains that he cannot bring himself to describe his father's death in the dungeon because 'certain things are not to be described. Lest we ease the burden of those who built them' (Chamoiseau 1997b, p. 36). Paradoxically, his silence signals not a willed forgetting or repression of the past, but rather an insistent remembrance of suffering.

Although Chamoiseau allows Esternome's silence to mark the impossibility of fully narrating Caribbean history, implicitly calling into question those historical accounts which posit themselves as truthful, *Texaco* insists on the attempt to recuperate the past despite its apparent inaccessibility. The slum-dwellers' impoverished existence may well be symptomatic of a seemingly 'impossible history', but nevertheless they experience the actual, lived effects of that traumatic legacy on a daily basis. More than mere symbols of suffering, the slum-dwellers can only overcome their contemporary marginalisation by articulating a history that restores their claim to a place in the city and on the island.

In Marie-Sophie, Chamoiseau locates the will to narrate history, despite its troubling content. Unlike her deliberately mute grandfather and her disturbed silent father, Marie-Sophie actively seeks language that will enable her to describe, comprehend and commemorate the past. Whereas Esternome is struck dumb by the extent of the devastation wrought by the volcano, Marie-Sophie insistently asks 'how does one mourn?' when she visits the ossuaries of Saint Pierre years later (Chamoiseau 1997b, p. 153). Her attempt to document Texaco's history is, in part, a response to this dilemma. As Samuel Durrant persuasively suggests, 'postcolonial narrative, structured by a tension between the oppressive memory of the past and the liberatory promise of the future, is necessarily involved in the work of mourning' (Durrant 2004, p. 1). Marie-Sophie's postcolonial history is, following Durrant's description, deliberately Janus-faced: she narrates the past in order to lay claim to a better future for her fellow slum-dwellers.

Postcolonial mourning poses a dilemma of both scale and substance. Faced with the piles of anonymous bones that literalise the weight of her ancestors' tragic past, Marie-Sophie struggles to come to terms with the extent of their loss. In asking how to mourn, she not only registers the considerable size of the task before her, but also an uncertainty about possible methods of remembrance. Although Pierre Nora (1989) suggests that archives, museums and other 'sites of memory' provide compensatory historical access in a modern era that has been irrevocably severed from the past, Marie-Sophie and her fellow slum-dwellers lack the space, time and means to construct such sites. Instead, Chamoiseau suggests that there are commemorative possibilities inherent in the Caribbean environment. Indeed, the landscape expresses trauma with an eloquence that escapes Esternome in the aftermath of the volcano. In later life, Esternome alerts Marie-Sophie to a tree that grows 'above Saint-Pierre's remains . . . a massive survivor of the volcano, spread like the spirit of a man who still owns his memories' (Chamoiseau 1997b, p. 118). Foreign to the island, the tree is an accidental transplant from Africa, which Congo slaves recognise and replant wherever possible. Although the ruins of Saint Pierre offer few monuments to the past, Chamoiseau demonstrates that the landscape itself asserts living memories.

Chamoiseau's recurrent tree motif further signals the inextricable trauma of the landscape and its inhabitants. When Esternome buries his mother, a tree similarly sprouts at her grave. She is killed by vengeful ex-slaves who overrun the plantation during the 1848 freedom riots. Her loyalty to the Béké is her downfall. When Esternome returns from Saint Pierre to find her body in a makeshift grave, he crafts a coffin from mahogany in which he reburies her. He refuses to allow her to be 'buried with-no-bell-a-tolling in guano sacks' (Chamoiseau 1997b, p. 104). By doing so, he not only postpones his mother's inevitable physical decay but, more importantly, he refutes her lifelong degradation and objectification. His insistence on providing her with 'some kind of funeral' asserts that she is not merely excrement. In an excerpt from Marie-Sophie's notes Esternome further observes that:

> [T]he red coffins shot up roots; and one could see several agony-trees, branches contorted with pain, rise on the backside of the years. Looking at them brought back memories one didn't have. It stiffened in you like a sad muffled drumbeat.
>
> (Chamoiseau 1997b, p. 105)

Although Esternome wishes to put his mother to rest, he overlooks the potential fertility of her remains. Like guano manure, her decomposing body literally nourishes the growth of trees whose twisted branches symbolically recall the tortured bodies of his uprooted ancestors, forcibly transported to the Caribbean as slaves. When Esternome sees these trees, he experiences the belated trauma associated with what Marianne Hirsch terms 'postmemory': 'traumatic recall at a generational remove' (Hirsch 2008, p. 106). The rhythmic resonance of his grief signals its African origins. As Hirsch suggests, inherited memories profoundly shape the imaginations of those whose parents witnessed collective trauma. In seismic fashion, the emotional aftershocks of slavery are deeply felt by subsequent generations. Lacking monumental fixity, the trees express living memories which cannot be buried nor suddenly healed.

Formal gleaning

If, as Esternome's response to the 'agony-trees' suggests, the Caribbean landscape offers an under-examined monument to the region's traumatic past, Chamoiseau makes it clear that the increasing urbanisation of the islands further obscures this embedded history. Aware of the urgent need to preserve the alternative archive offered by her parents' memories, Marie Sophie conscientiously records them in notebooks that she appropriately salvages from trash. Interestingly, the retrieval of an old accounting ledger catalyses her transcription of Texaco's oral history. A different economy is at work in her narration. By recording her parents' words in these cast-off notebooks, she demonstrates a respect for both material and discursive *rejectamenta*. Her historical narrative revalues the stories, people and things deemed unprofitable by the colonial system and its neocolonial form. Marie-Sophie is not interested in individual gain, but the production of a communal resource that will secure Texaco's standing in the city.

Marie-Sophie's notebooks inevitably invoke the poetic *cahier* of her famous compatriot, Aimé Césaire, who features in the novel as both an inspirational politician and a reluctant benefactor. Following his personal visit to the slum, the 'Césaire effect' ensures that Texaco is protected from further police harassment (Chamoiseau 1997b, p. 354). Despite her evident admiration for 'Papa Césaire', Marie-Sophie's notebooks qualify the spirit of resistance inscribed in his *Cahier d'un retour au pays natal* (Césaire, 1939). Whereas Césaire stridently celebrates the African roots of the island's diasporic population, Marie-Sophie is concerned with the remembrance of a specifically Martinican identity. In contrast to Césaire's unidirectional 'return' to Africa, her history pulls in many directions, drawing on multiple pasts in order to secure a future in which Texaco will be recognised as a legitimate part of the city and the island's culture as a whole. Whereas, Chamoiseau and his co-authors argue in *Eloge de la Créolité*, Césaire 'restored mother Africa, matrix Africa, the black civilization' to Caribbean culture, Marie-Sophie is attentive to what Stuart Hall terms the multiple 'présences' within Martinican identity: not just African, but also European and American (Bernabé *et al.* 1993, p. 79; Hall 1994).[16]

Marie-Sophie is concerned that the conversion of her late father's oral history into written French will arrest the necessary multidirectionality of the history he passed on to her. Her transcription of his recollections into 'immobile notebooks' renders them disembodied, as if she 'were burying him again' (Chamoiseau 1997b, p. 321). Concerned about preserving the immediacy of his words, Marie-Sophie wonders:

> [I]s there such a thing as writing informed by the word, and by the silences, and which remains a living thing, moving in a circle, and wandering all the time, and which reinvents the circle each time like a spiral which at any moment is in the future, ahead, each loop modifying the other, nonstop, without losing a unity difficult to put into words?
>
> (Chamoiseau 1997b, p. 322)

Marie-Sophie's doubts about the efficacy of written narration not only speak to her desire to preserve her father's vitality, but also to Caruth's concern about the difficulty of representing a traumatic history without diminishing its affective impact. As Sam Durrant explains, in the face of senseless, overwhelming traumas, such as slavery and the Holocaust, 'postcolonial narrative is confronted with the impossible task of finding a mode of writing that would not immediately transform formlessness into form, a mode of writing that can bear witness to its own incapacity to recover a history' (Durrant 2004, p. 6).

Chamoiseau's text embodies some of the unpredictability sought by Marie-Sophie. According to the afterword, the novel is a modified transcription of the oral history Marie-Sophie provides when he meets her during his research on the storyteller Solibo Magnifique, the subject of his previous novel. Intrigued by the matadora's 'profound authority', Chamoiseau decides to compile her fragmented history of Texaco. In addition to the inevitable distance created by translating her spoken word into written form, the author explains that Marie-Sophie's account has been further mediated by her inconsistent powers of recall and his 'bastard of a tape recorder' that did not capture everything she said (Chamoiseau 1997b, p. 387). Additionally, the Creole-speaking Marie-Sophie asks him to '"fix up" her speech into good French' (Chamoiseau 1997b, p. 388). Marie-Sophie's narrative thus reaches the reader in multiply mediated form. Throughout his adjusted transcription of Marie-Sophie's words, Chamoiseau interweaves excerpts from the notebooks in which she earlier recorded her parents' memories. Although he carefully preserves these notebooks – he numbers them, repairs them and stores them in the Schoelcher Library – Chamoiseau does not present them chronologically in the body of the text, drawing attention to his distinctive archival practice in *Texaco*.

Many critics have noted that the resulting composite text exemplifies artistic *bricolage*, a practice first defined by Lévi-Strauss as the art of 'making do with "whatever is at hand", that is to say with a set of tools and materials which is always finite and is also heterogeneous' (Lévi-Strauss 1966, p. 17). Maeve McCusker points out the correspondence between Chamoiseau's authorial strategy and the building techniques adopted by Esternome in his youth when he begins to make a living as a real-life *bricoleur*-handyman (McCusker, 2007). Following the example of his mentor Théodorus Sweetmeat, he learns how to add 'Norman knowledge to the teachings offered by the African huts and Carib longhouses' in order to construct houses uniquely suited to the island environment and available materials (Chamoiseau 1997b, p. 57). As McCusker observes, 'the text becomes itself a sort of building site, drawing attention to its multiple materials rather than to any sense of cohesion or coherence. It is the disparate and interweaving threads binding the text(ile)/*Texaco* together which are privileged over any sense of seamlessness' (McCusker 2007, p. 110). In this view, *bricolage* defiantly celebrates the region's fragmented origins, despite ongoing hardships.

At its best, aesthetic *bricolage* is an innovative, adaptive process that captures the cultural creolisation or mixture that characterises the Caribbean. However, in a recent article, Wendy Knepper calls for a renewed understanding of *bricolage*, which is attentive to the political motivations of such postcolonial *bricoleurs* as Chamoiseau. Following Lévi-Strauss's originally ahistorical conception, she notes that the all-too-easy celebration of *bricolage* can mask 'possibilities for cultural impoverishment as a result of the deliberate obliteration or unconscious repression of cultural fragments' (Knepper 2006, p. 72). Although she admires Chamoiseau's strategic *bricolage*, which serves as 'not only a form of counter-cultural creativity but also as a critique: a radical ideological and imaginative reconstruction of society, at once blasting apart and reframing the context of interpretation by placing the known fragments in new arrangements' (Knepper 2006, p. 82), Knepper highlights the uneven power dynamics involved in the creation of attractively 'hybrid' artistic products. In a recent analysis of the ways in which European modernism inflects and is influenced by postcolonial poetics, Jahan Ramazani (2006) tellingly replaces the term '*bricolage*' altogether, preferring the term 'postcolonial hybridity' to describe the latter's aesthetic engagement with cross-cultural experience.

In its reference to the process of artistic composition, gleaning offers a complementary critical concept to these qualified notions of *bricolage*. Reading Chamoiseau as both *bricoleur* and *glaneur* allows for a fuller understanding of the acts of observation, selection and retrieval that contribute to his creole art, rather than limiting attention to their aesthetic outcome. Attention to his gleaning grounds Chamoiseau's alternative bricolage aesthetics in the particular context of the historically uneven modes of production from which they emerge. His composite textual format is further illuminated by Ursula Le Guin's 'Carrier Bag theory of fiction', which rejects the 'linear, progressive, Time's-(killing)-arrow mode of the Techno-Heroic' with which the realist novel is associated, recasting the novel as a cultural 'container', the purpose of which 'is neither resolution nor stasis but continuing process' (Le Guin 1996, p. 153). Drawing on a playful stone-age analogy, in which the author is figured as a female gatherer (as opposed to a male hunter), she aptly proposes that 'the proper, fitting shape of the novel might be that of a sack, a bag...a medicine bundle, holding things in a particular, powerful relation to one another and to us' (Le Guin 1996, p. 153). Chamoiseau, as authorial gleaner of Fort-de-France's history, crafts just such a cultural container in *Texaco*. By combining the various strands of Texaco's long-overlooked past into a single narrative, he models a promising method of polyphonic historical narration that, contrary to the principle of cultural homogenisation on which French assimilation operates, respects the 'open specificity' of the Caribbean (Bernabé *et al.* 1993, p. 89).

Conclusion

Despite the critical promise of Chamoiseau's formal gleaning, it should not be forgotten that the material practice that he creatively appropriates emerges from conditions of undesirable hardship and inequity. As a means of opposition, such micro-strategies are limited in their capacity to achieve far-reaching social change, as revealed by their attempted formalisation. In a recent analysis of foraging practices within contemporary American cities, Michelle Coyne (2009) examines efforts to systematise and diffuse the philosophy of freeganism, which advocates the recuperation of discarded food in a manner akin to the physical gleaning Marie-Sophie conducts. As Coyne points out, such practices are dependent on the social structure that they simultaneously challenge: 'It is clear that the waste system that horrifies the Freegan has also become the sustenance upon which she depends' (Coyne 2009, p. 14). Gleaning is always dependent on the very structures that it seeks to contest. Even if gleaning reveals local and transnational histories that productively denaturalise the social conditions that lead to its necessity, it still always results in an asymmetrical symbiosis between the needy and the excessive.

Chamoiseau's formal gleaning is thus deeply ironic. His stylistic excess is enabled by the very conditions of deprivation that his subjects experience. His painstaking compilation of Texaco's history simultaneously highlights the cultural significance of the slum and marks its material and social degradation. However, the ironic gap between Chamoiseau's form and content – between textual excess and material lack – usefully reasserts the distance between figurative waste and actual waste. This irony is deepened when one considers the differing effects of Marie-Sophie's oral history and Chamoiseau's textual account. On the one hand, Marie-Sophie's narrative ultimately results in Texaco's incorporation into Fort-de-France: by detailing the slum-dwellers' formative role in the emergence of the city, starting with its plantation origins, she asserts a persuasive claim to full local and global citizenship for herself and her peers. Chamoiseau's gleaned narrative of the slum, by contrast, insists on its distinctiveness by highlighting the unique creativity and resilience of Texaco. As narrative method and interpretive tool, gleaning thus refuses to naturalise or legitimate urban marginalisation.

The downside to the slum's incorporation, as Chamoiseau notes in his afterword, is the taming of Texaco's resistant energy, the critical force that emerges, with painful irony, from precisely those conditions of deprivation that he seeks to alleviate. The slum's absorption into the city risks new historical amnesia as the traces of Texaco's past are concealed by material developments such as the improvement of the shacks, and the gradual forgetting of younger Texaco generations. Chamoiseau's nuanced gleaning, and the narrative tension it produces, thus leaves the reader with the enduring problem of the urban margins: how can they be fully acknowledged and accommodated if not through their absorption into the same system that created them?

The following chapter examines a novel that portrays the violent elimination of slums as the unwelcome alternative to their assimilation. Set in 1980s Lagos, Chris Abani's *GraceLand* (2004) dramatises the state-sanctioned demolition of a densely populated shantytown. While his text expresses a deeper pessimism about the potential for state recognition of slum-dwellers' needs and rights, Abani shares with Chamoiseau an insistence on the restorative potential of the imaginative arts, which he reveals to be criminally censored in late twentieth-century Nigeria.

Notes

1 See 'Un reportage sur "les békés"' (2009) and 'France Faces Unrest in Caribbean' (2009).

2 See Fidler (2009), Nouvet (2009).

3 See Armand Nicolas's *Histoire de la Martinique* (1996) for a comprehensive history of the island from the early seventeenth century to 1971. Laurent Jalabert's *La Colonisation sans nom* (2007) analyses the island's neocolonial relationship to France from 1960 to the present.

4 The particularly ambiguous relationship between Martinique and France perhaps explains why so many significant authors of postcolonial critique have emerged from this relatively small island, amongst them Aimé Césaire, Frantz Fanon and Edouard Glissant.

5 See Jamaica Kincaid's *A Small Place* (1988) for an especially incisive critique of the complacent historical amnesia enjoyed by foreign visitors to her home island of Antigua.

6 Chamoiseau's commitment to urban storytelling is distinctive. In his first autobiographical volume, he recalls that 'city storytellers were rare' when he was young. His own fictional work infuses Fort-de-France with the same imaginative richness that he remembers hearing in the 'Creole tales from the country' brought to his childhood home by Jeanne-Yvette, a compelling 'real storyteller' (Chamoiseau 1999, p. 70).

7   See, for example, C. L. R. James, *Minty Alley* (1936); Joseph Zobel, *La Rue Cases-Nègres* (1950); Orlando Patterson, *The Children of* Sisyphus (1964); and Earl Lovelace, *The Dragon Can't Dance* (1979).

8   Derived from the Creole *matadò* meaning 'one who triumphs . . . like a matador in the arena', Chamoiseau invents the French term *femme-matador* to convey that Marie-Sophie is 'a strong, respected, authoritative woman' (Chamoiseau 1997b, p. 400).

9   For additional reviews, see Phillips (1997), Michaels (1997), Scott Fox (1997).

10 Of course, the novel's reception was not universally positive. The French literary establishment's welcoming embrace of a self-professed anticolonial activist writer was an irony that did not go unremarked by some. In a cooler appraisal of the novel, Thomas C. Spear suggests that Chamoiseau's 'mindlessly heavy' linguistic turns served the commercial intentions of his French publisher, rather than the entertainment of the reader (Spear 1993, p. 157).

11 See Milne (2006), McCusker (2007), Chancé (2010).

12 The biblical Ruth is the archetypal gleaner – displaced from her homeland, she provides for herself by gleaning in fields owned by Boaz, whom she later marries. In Keats's 'Ode to a Nightingale,' the speaker recalls her solitary foraging as he ponders whether the bird's song also 'found a path / Through the sad heart of Ruth, when, sick for home, / She stood in tears amid the alien corn'.

13 See Peter King (1991) for a brief history.

14 In *Childhood*, Chamoiseau shows an awareness of Millet's work as he remembers how the Syrian shopkeepers in the Fort-de-France of his youth exploited 'country people['s] . . . soft spot for Millet's *Angelus*' in order to attract them to their stores (Chamoiseau 1999, p. 78). This wry recollection sounds a note of caution against idealising the humble peasantry depicted in this painting. The gleaning that Marie-Sophie and her urban peers undertake has a long history, but it is not a nostalgic practice.

15 See Pieter Hugo, *Permanent Error* (2011); also Andrew McConnell's 'Rubbish Dump 2.0' (2010) and Kevin McElvaney's 'Agbogbloshie: The World's Largest E-waste Dump' (2014).

16 Chamoiseau draws on Hall's terminology in a recent interview in which he explains his understanding of Martinique's distinctively creole culture: 'For my part, I believe more in what I call "presences", and I think we are coming out of the era of closed territories and entering the era of "places". The place, then, is multi-transcultural, multi-translingual; the place is inhabited by diversities and so, in that sense, there aren't any diasporas because the diverse, the "other" is already there – everywhere. So, what you have are emanations, networks, and presences. The place of Martinique, for example, has an African presence, an American presence, a European presence across the history that it had with France – and all that constitutes the networks of solidarity and presences that sustain Martinique as a place' (Morgan 2008, pp. 448–49).

References

Abani, C. (2004) *GraceLand*. New York: Picador.

Assman, A. (2002) 'Beyond the Archive', in Neville, B. and Villeneuve, J. (eds.) *Waste-site Stories: The Recycling of Memory*, Albany, NY: SUNY Press, pp. 71–83.

Bernabé, J., Chamoiseau, P. and Confiant R. (1993 [1989]) Éloge de la Créolité (trans. M. B. Taleb-Khyar), Paris: Gallimard.

Caruth, C. (1995) *Trauma: Explorations in Memory*. Baltimore, MD: Johns Hopkins UP.

Césaire, A. (2000 [1939]) *Cahier d'un retour au pays natal*, Columbus, OH: Ohio State UP.

Chamoiseau, P. (1997a) Écrire en pays dominé. Paris: Gallimard.

Chamoiseau, P. (1997b [1992]) *Texaco* (trans. from French and Creole by R-M Réjouis and V. Vinokurov), New York: Vintage.

Chamoiseau, P. (1999 [1993]) *Childhood* (trans. from French by C. Volk), Lincoln, NE: U of Nebraska P.

Chancé, D. (2010) *Patrick Chamoiseau, Écrivain Postcolonial Et Baroque*. Paris: Champion.

Coyne, M. (2009) 'From Production to Destruction to Recovery: Freeganism's Redefinition of Food Value and Circulation', *Iowa Journal of Cultural Studies*, vols. 10/11: pp. 9–24.

Crummy, I. (1999) 'The Subversion of Gleaning in Balzac's *Les Paysans* and in Millet's "Les Glaneuses"', *Neohelicon*, vol. 26, no. 1, pp. 9–18.

Dawson, A. (2004) 'Squatters, Space, and Belonging in the Underdeveloped City', *Social Text*, vol. 22, no. 4, pp. 17–34.

Durrant, S. (2004) *Postcolonial Narrative and the Work of Mourning: J. M. Coetzee, Wilson Harris, and Toni Morrison*. Albany, NY: SUNY Press.

Fidler, R. (2009) 'Martinique General Strike Ends in Victory', *Links: International Journal of Socialist Renewal*. 18 March [Online] Available at http://links.org.au/node/956 (Accessed 30 December 2015).

'France Faces Unrest in Caribbean' (2009), *BBC News*, 12 February [Online]. Available at http://news.bbc.co.uk/1/hi/world/europe/7885683.stm (Accessed 30 December 2015).

Gallagher, M. (2002) *Soundings in French Caribbean Writing Since 1950: The Shock of Space and Time*, London: Oxford UP.

Glissant, E. (1989 [1981]) *Caribbean Discourse: Selected Essays* (trans. from French by J. M. Dash), Charlottesville, VA: U of Virginia P.

Gosson, R. (2006) 'What Lies Beneath? Cultural Excavation in Neocolonial Martinique', in Hood Washington, S., Rosier, P. C. and Goodall, H. (eds.) *Echoes from the Poisoned Well: Global Memories of Environmental Injustice*, Lanham, MD: Rowman & Littlefield. pp. 225–43.

Grabski, J. (2009) 'Urban Claims and Visual Sources in the Making of Dakar's Art World City', *Art Journal*, vol. 68, no. 1, pp. 6–23.

Hall, S. (1994) 'Cultural Identity and Diaspora', in Williams, P. and Chrisman, L. (eds.), *Colonial Discourse & Postcolonial Theory: A Reader*, New York: Columbia UP. pp. 392–403.

Haraway, D. (2009) 'Service, Charity, Poverty, and the Right to Food', *Gleaning Stories, Gleaning Change* [Online]. Available at http://humweb.ucsc.edu/gleaningstories/html/gleaners/donna.html (Accessed 30 December 2015).

Hirsch, M. (2008) 'The Generation of Memory', *Poetics Today*, vol. 29, no. 1, pp. 103–28.

Hugo, P. (2011) *Permanent Error*, Munich: Prestel Art.

Jalabert, L. (2007) *La Colonisation sans nom: La Martinique de 1960 à nos jours*. Paris: Indes Savantes.

James, C. L. R. (1936) *Minty Alley*. London: Secker & Warburg.

Kemedijo, C. (2001) 'De *Ville Cruelle* de Mongo Beti à *Texaco* de Patrick Chamoiseau: Fortification, Ethnicité et Globalisation dans la Ville Postcoloniale', *L'Esprit Créateur*, vol. 41, no. 3, pp. 136–50.

Kincaid, J. (1988) *A Small Place*. New York: Penguin.

King, A. (1984). 'Colonial Architecture Re-visited: Some Issues for Further Debate', in Ballhatchet, K. and Taylor, D. (eds.) *Changing South Asia: City and Culture*. London: U of London, pp. 99–106.

King, P. (1991). 'Customary Rights and Women's Earnings: the Importance of Gleaning to the Rural Labouring Poor, 1750–1850', *Economic History Review*, vol. 44, no. 3, pp. 461–76.

Knepper, W. (2006) 'Colonization, Creolization, and Globalization: The Art and Ruses of *Bricolage*', *Small Axe*, vol. 11, no. 1, pp. 70–86.

Le Guin, U. K. (1996) 'The Carrier Bag Theory of Fiction', in Glotfelty, C. and Fromm, H. (eds.), *The Ecocriticism Reader: Landmarks in Literary Ecology*, Athens, GA: U of Georgia P, pp. 149–154.

'Les Derniers Maîtres de la Martinique' (2009), *Spécial Investigation*, Canal+, 6 February.

Les Glaneurs Et La Glaneuse (2002) Directed by Agnès Varda [DVD]. New York: Zeitgeist Video.

Lévi-Strauss, C. (1966 [1962]) *The Savage Mind*. Chicago, IL: U of Chicago P.

Lionnet, F. and Shi, S. (2005) *Minor Transnationalism*. Durham, NC: Duke UP.

Lovelace, E. (1979) *The Dragon Can't Dance*. Harlow: Longman.

McConnell, A. (2010) 'Rubbish Dump 2.0', *Andrew McConnell* [Online]. Available at http://www.andrewmcconnell.com/Rubbish-Dump-2.0/1 (Accessed 30 December 2015).

McCusker, M. (2007) *Patrick Chamoiseau: Recovering Memory*. Liverpool: Liverpool UP.

McElvaney, K. (2014) 'Agbogbloshie: The World's Largest E-Waste Dump', *The Guardian*, 27 February [Online]. Available at www.theguardian.com/environment/gallery/2014/feb/27/agbogbloshie-worlds-largest-e-waste-dump-in-pictures (Accessed 30 December 2015).

Michaels, L. (1997) 'Mother Tongues', *New York Times*, 30 March [Online]. Available at https://www.nytimes.com/books/97/03/30/reviews/970330.30michaet.html (Accessed 30 December 2015).

Milne, L. (2006) *Patrick Chamoiseau: Espaces d'une écriture antillaise*. Amsterdam: Rodopi.

Morgan, J. (2008) 'Re-Imagining Diversity and Connection in the Chaos World: An Interview with Patrick Chamoiseau', *Callaloo*, vol. 31, no.2, pp. 443–53.

Nicolas, A. (1996) *Histoire de la Martinique*. 3 vols. Paris: L'Harmattan.

Nora, P. (1989) 'Between Memory and History: Les Lieux de Mémoire', *Representations*, vol. 26, pp. 7–25.

Nouvet, F. (2009) 'En Martinique, c'est toujours "le combat des mêmes contre les memes"', *L'Humanité*, 29 July {Online]. Available at http://www.humanite.fr/node/421315 (Accessed 30 December 2015).

Patterson, O. (1964) *The Children of Sisyphus*. Boston, MA: Houghton Mifflin.

Phillips, C. (1997) 'Unmarooned', *New Republic*, vol. 216, no. 17, pp. 3–4.

Quartey, K. (2011) *Children of the Street*. New York: Random House.

Ramazani, J. (2006) 'Modernist Bricolage, Postcolonial Hybridity', *Modernism/Modernity*, vol. 13, no. 3, pp. 445–63.

Rubenstein, M. (2008) 'Light Reading: Public Utility, Urban Fiction, and Human Rights', *Social Text*, vol. 26, no. 4, pp. 31–50.

Scott Fox, L. (1997) 'Like Heaven', *London Review of Books*, 22 May: pp. 18–19.

Spear, T. C. (1993) *Texaco* by Chamoiseau, P. Reviewed in: *The French Review*, vol. 67, no.1, pp. 157–58.

'Un reportage sur "les békés" enflamme la Martinique' (2009), *Le Monde*, 13 February [Online]. Available at www.lemonde.fr/politique/article/2009/02/13/un-reportage-sur-les-bekes-enflamme-la-martinique_1154769_823448.html (Accessed 30 December 2015).

Walcott, D. (1997) 'A Letter to Chamoiseau', *The New York Review of Books,* 14 August [Online]. Available at www.nybooks.com/articles/1997/08/14/a-letter-to-chamoiseau/ (Accessed 30 December 2015).

Young, R. (2001) *Postcolonialism: An Historical Introduction*. Oxford: Blackwell.

Zobel, J. (1950) *La Rue Cases-Nègres*. Paris: Froissart.

2 'Suspended city'

Personal, urban and national development in Chris Abani's *GraceLand*

Addressing the Africa Leadership Forum in 1988, Colonel Raji Rasaki, Military Governor Of Lagos State prefaced his lengthy recommendations for the better management of the then Nigerian capital with the assertion that:

> Metropolitan Lagos means so many different things to its diverse inhabitants and visitors. To some, it is the centre of civilisation, sophistication, wealth, opulence and the haven of the elite. To others, it is the heart of decadence where only the fittest survive, a jungle city of chaos where nothing works but for pickpockets, armed robbers and fraudulent characters.
>
> (Rasaki 1988, p. 3)

Highlighting the different perceptions of the city in his charge, the governor's comments suggest the contrasting urban experiences of those who visit, live and work in Lagos; experiences which have only become more divergent in recent years as Lagos has grown to a mega-city of more than 13 million people, belying its sixteenth-century origins as a small fishing and farming village.[1] This staggering demographic and territorial expansion has produced both conspicuous wealth in such affluent neighbourhoods as Victoria Island, together with the proliferation of slums in and around the city, the inhabitants of which bear the brunt of Lagos's manifold environmental, infrastructural and economic problems.[2]

The city's continued increase in size and population despite an economic downturn at the urban and national level has confounded expected trajectories of development, prompting some external commentators to suggest that there is something inherent to sub-Saharan African cities that forestalls their capacity to spark economic growth.[3] While critics argue that such measures of development are implicated in neocolonial discourses of modernisation that perpetuate Western dominance of a perpetually 'developing' world, the postcolonial state in Nigeria and elsewhere has frequently invoked the burden of national progress in order to pass off development initiatives that will primarily benefit an elite minority as matters of broad public interest.[4] This is especially true in Lagos, the urban landscape of which functions as a key symbol of national pride. In his address, Colonel Rasaki goes on to lament the city residents' apparent lack of cooperation with the development efforts of government during his tenure as 'action governor', a nickname earned through his advocacy of modernisation programmes:

> To those charged with the management of the city, [Lagos] is inhabited by an articulate and fastidious citizenry quick to criticise, expecting everything free from Government, yet compounding, day-by-day, the problems that the Government faces in its efforts to improve the lives of the majority.
>
> (Rasaki 1988, p. 3)

Although Lagos is, as poet Odia Ofeimun puts it, 'the closest Nigerian parallel to a melting pot . . . our prime city of crossed boundaries' (Ofeimun 2001, p. 138), it is also a site of entrenched social, political and economic division. As Martin Murray and Garth Myers note in their critical analysis of urban planning initiatives similar to Rasaki's, such 'efforts to create and maintain orderly urban landscapes are inextricably linked with the process of *boundary-making*' (Murray and Myers 2006, p. 237; emphasis added). By portraying the government's attempts at civic improvement as valiant efforts in the face of intransigent local obstacles, Rasaki implies a clear divide between the citizens of Lagos and the state authorities. His invocation of paternalistic colonial discourse is indicative of a lingering top-down approach to development within the postcolonial state that endures today, making it hard to find solutions to the problems that particularly impact marginalised populations.

This chapter examines a novel that continues a long tradition of Lagos literature in its creative engagement with the immense scale and elusive substance of the city. In *GraceLand* (Abani 2004), Chris Abani employs a largely realist style that shares both descriptive power and occasional ironic humour with such early precedents as Cyprian Ekwensi's *People of the City* (1954) and Chinua Achebe's *No Longer at Ease* (1960). Like the protagonists of their respective novels, Abani's central character is a young male migrant to Lagos who struggles to navigate the financial demands, family expectations and personal aspirations that come with living in the city. As a marginalised slum-dweller, however, his socio-economic status is more precarious than that of Ekwensi's Sango and Achebe's Obi, both of whom have access to relatively stable housing and employment. By re-imagining the city from the perspective of its poorest population, Abani challenges elitist responses to Lagos's social and infrastructural problems. In his specific focus on the slum of Maroko, he joins a number of Nigerian writers who have chosen to represent this district in fiction, poetry and drama; amongst them Wole Soyinka, J.P. Clark and Maik Nwosu.[5] This Maroko corpus is broadly concerned with describing and critiquing the forced eviction of hundreds of thousands of people from their homes in what Abani refers to throughout the novel as a 'ghetto' adjacent to affluent Victoria Island.[6] Although the area was decisively cleared under Colonel Rasaki's direction in 1990, Abani's novel returns us to an earlier slum clearance in 1983 – a violent prelude to the more comprehensive demolition programme that was to follow.[7]

Abani's extended engagement with the vagaries of slum existence creates a suggestive metonym for the instability and inconsistency of the postcolonial nation-state, allowing him to effectively diagnose and critique the causes of the disparate living conditions of Lagos's numerous residents. Specifically, through his formal and thematic elaboration of a 'suspension' leitmotif, Abani demonstrates the paralysing imbrication of local, national and international discourses of development, especially for those who live at the urban margins. By invoking and subverting the form of the *Bildungsroman*, Abani exposes the discrepant trajectories of development that exist within a single city, suggesting the untenability of national models of development which perpetuate a 'First World'/'Third World' hierarchy. Through his narration of the experiences of a young slum-dweller who fails to meet the *Bildungsroman*'s generic expectations of 'formation', 'education' and 'coming-of-age', Abani further demonstrates that both state and society inhibit progress in their frequently violent promotion of inflexible measures of development. While the government's brutal policing of their status quo is clearly detrimental to social cohesion, Abani suggests that a reactionary recourse to invented traditions in the face of unwelcome modernisation is equally inimical. Subject to the varying expectations of his father, his friends, his neighbours and his fellow Nigerians, his protagonist Elvis lives a suspended existence characterised by alienation from his urban surroundings, estrangement from his family and an inability to take action against people and practices he knows to be wrong. Not only is he a paralysed observer of social injustice, but his body also bears scars of abuse, exploitation and self-loathing that inhibit his psychic and physical maturation into adulthood. Although Elvis finds some solace in transnational cultural exchange, this is circumscribed by his simultaneous immersion in a global economic system that perpetuates his marginalisation. In his fictional representation of Lagos's uneven development, Abani thus responds to calls for more holistic accounts of development than those which emerge from other 'texts of development' such as urban plans, economic reports, political speeches and statistical analyses (Crush 1995, p. 5).[8]

GraceLand as critical *bildungsroman*

GraceLand narrates the troubled coming of age of Elvis Oke. Born in Afikpo, 'a dusty end-of-the-highway fishing town' in southeastern Nigeria in 1967, the relative comfort of Elvis' childhood is overshadowed by his mother Beatrice's long battle with breast cancer (Abani 2004, p. 146). Despite hospital treatments and the application of his grandmother's homemade herbal remedies, she dies when Elvis is eight years old. His father Sunday seeks consolation in alcohol, becoming ever more estranged from his son, whose burgeoning interest in dancing and performance enrages his masculinist sensibilities. When the military government announces its intention to step down in 1980, Sunday resigns his secure position as superintendent of schools and decides to run for office in the new civilian government, having served as a member of parliament in the first republic during the 1960s. His failed election campaign bankrupts him, forcing Elvis and Sunday to move to Lagos in search of work. They find a home, but scarce employment in Maroko, one of the city's many slums. As his father's alcoholism worsens, Elvis ekes out an existence by impersonating his American namesake for tourists on the beaches of Victoria Island. When his financial circumstances become untenable, his friend Redemption, a self-professed 'original area boy' (Abani 2004, p. 55), takes him under his wing and introduces him to some of the more lucrative, and less legal, occupations that Lagos offers. Elvis tries his hand at escorting rich women to nightclubs and wrapping cocaine 'for export' (Abani 2004, p. 110), but he balks at smuggling kidnapped children and human body parts out of the country. His conscientious objection attracts the unwelcome attention of the Colonel, the corrupt chief of security to the head of state who had commissioned Redemption to assist with this job. Fearing fatal recrimination, Elvis flees Lagos with the Joking Jaguars, a dance troupe led by the Beggar King, an enigmatic vagrant who befriended Elvis during his early wanderings in the city. Meanwhile, Maroko is slated for demolition during the government-sponsored 'Operation Clean the Nation'. Sunday heads the slum-dwellers' protest against the destruction of their homes, but he is crushed by a bulldozer when he refuses to leave his shack. The novel concludes with the orphaned Elvis poised to emigrate to America, using a forged passport that Redemption gave him, having sacrificed his own dream of leaving the country.

As this overview of the novel suggests, *GraceLand* shares many plot features commonly associated with the *Bildungsroman*: 'childhood, the conflict of generations, provinciality, the larger society, self-education, alienation, ordeal by love, the search for a vocation and a working philosophy' (Buckley 1974, p. 18). Although the generic instability of the *Bildungsroman* has prompted much critical debate, with Marc Redfield claiming that it is a 'phantom formation' which 'exemplifies the ideological construction of literature by criticism' (1996, p. vii), Franco Moretti identifies this flexibility, or 'morphological *bricolage* and ideological compromise', as precisely the feature which enabled the *Bildungsroman* to flourish from the late eighteenth to the mid-nineteenth century in Europe (Moretti 2000, p. xii). Following Lukàcs's earlier influential definition of the genre's theme as 'the reconciliation of the problematic individual, guided by his lived experience of the ideal, with concrete social reality' (Lukàcs 1971, p. 132), Moretti identifies the source of this subjective challenge with the onset of modernity. In keeping with Bakhtin's assertion that the genre is unique in its depiction of a protagonist who 'emerges *along with the world* and reflects the historical emergence of the world itself' (Bakhtin 1986, p. 23), Moretti (2000) argues that, in the *Bildungsroman*, the transformations of youth come to stand for the challenges of modernisation. Crucially, the inevitable ending of youth brings narrative closure to social changes that are otherwise deeply unsettling.

The bourgeois European *Bildungsroman* that Moretti describes confers a seemingly natural progressive temporality on social changes that produced distinctly uneven spatial formations and discrepant environmental impacts. In his pertinent reading of the colonial *Bildungsroman*, Joshua D. Esty argues that the revised spatial imaginary brought about by imperial expansion challenges the form's ideology of synchronous progress. Highlighting what he considers to be a blindspot in Moretti's argument, namely that it is *'nationhood* [which] supplies the bildungsroman with a language of historical stability, a final form amidst the vast changes of industrialisation' (Esty 2007, p. 413; emphasis added), he argues that the uneven processes of global modernisation fostered by European colonialism call this 'cultural mechanism' into question (Moretti 2000, p. 9). His intervention usefully suggests the temporal contradiction inherent in colonial modernity, which demands development at the imperial 'centre' while the assumed 'periphery' must remain in a constant state of becoming in order for this power relation to be maintained.

If European imperialism destabilised the national borders that had previously lent a reassuring structure to the unprecedented social, industrial and cultural transformations induced by modernisation, *GraceLand* extends Esty's analysis by highlighting the stark unevenness of development that persists within the postcolonial state. Abani's attention to Lagos enables him to fully engage with the multiple centres and margins that exist within Nigeria, given the city's particular entanglement in criss-crossing networks of transnational and local power. Although some have questioned whether the *Bildungsroman*, so closely affiliated with a modern European bourgeois sensibility, can adequately represent postcolonial reality, Abani joins such African writers as Ngũgĩ wa Thiong'o and Tsitsi Dangarembga in crafting what Joseph Slaughter terms a 'dissensual *Bildungsroman*' that 'puts in circulation a countercultural narrative that seeks to rearticulate the sociohistorical universal and the political terms of antagonism between a marginal group and the dominant group' (Slaughter 2007, p. 184).[9] By appropriating the genre in order to narrate a story of stalled urban development, Abani calls the myth of unified national development into question.

At the level of form in particular, the novel's narrative structure demonstrates the multiple temporalities which intersect the space of the Nigerian nation. The novel is divided into two books, the first of which alternates between chapters narrating Elvis' childhood in Afikpo during the 1970s and chapters which record a single year of his new life in Lagos. Although there are no flashbacks in the second book, the geography of the chapters alternates between Lagos and the various other places that Elvis visits during his adventures with Redemption and the Beggar King. The effect of these temporal and spatial inconsistencies is twofold. From the very beginning of the novel, Abani uses chapter headings to clearly and deliberately invoke a specific period in Nigeria's history. *GraceLand* opens in Lagos in 1983 at a time when the country was on the verge of another extended period of military rule. Unlike other contemporary Nigerian authors who have chosen to explore Nigeria's recent past,[10] Abani returns the reader to a moment prior to the tyranny of the Abacha regime from 1993–98 when the full force of economic depression and government corruption had yet to be felt.[11] Although Abani situates his novel in a specific historical context, his concern is with the day-to-day impact of major events on those whose histories are often sidelined. By focalising the country's political shifts and social upheavals through the young, curious Elvis, he provides a destabilising alternative to the implied dominant history that shadows the actual story he tells. For example, when the military government agrees to make way for the civilian Second Republic in 1980, Elvis' sceptical reaction undermines fidelity to a single historical account. He learns of this significant political transition from the newspaper that his father is reading:

The headline caught Elvis' attention: MILITARY TO STEP DOWN. That was strange; Elvis could not remember when the military had not run the country. His father spoke often and nostalgically about his days as a member of parliament in the first republic, but to Elvis it sounded suspiciously like all his father's stories. Like the one about being made to walk forty miles each way to school every day as a child. Or the one about hunting a lion with his father, Elvis' grandfather, armed with nothing but native broadswords. Of course, his father did not know that in general science, Elvis had learned that lions had been extinct in this part of the country since the twenties.

(Abani 2004, p. 174)

By aligning the newspaper's official, reported history with the personal myth-making of his father, Elvis humorously and unwittingly indicates the competition for narrative dominance on which any historical record is contingent. Benedict Anderson has argued that newspapers, like other products of print capitalism such as novels, are crucial to the formation of unified national identity because the daily 'mass ceremony' of reading the news provides a common material focus for the disparate subjects of an 'imagined community', the extent and content of which is otherwise difficult to grasp (Anderson 1991, p. 35). However, Elvis' subjective reaction to the headline in his father's paper resists his interpellation into the national collective, suggesting the instability of its temporal and geographical boundaries. As Patrick Chamoiseau does in *Texaco*, Abani seems to be concerned with pluralising history. In doing so, he fragments the 'image of *man growing* in *national–historical time'* which Bakhtin identified as the central mechanism of the *Bildungsroman* (Bakhtin 1986, p. 25). Although Jonathan Crush identifies 'the national development plan' as 'the basic liturgy of post-World War II development discourse', Abani's deconstruction of unified national time suggests the need for alternative development frameworks that take into account the micro-scale of marginalised Nigerian existence (Crush 1995, p. 8–9).

Abani also calls the integrity of the nation into question by narrating Elvis' time in Lagos at a slower pace than his rural childhood. This explicit formal challenge to what Esty terms 'the *bildungsroman*'s basic genetic code of progressive temporality' conveys a sense of the past catching up with the present (Esty 2007, pp. 425–26). While the reader notes the passing of earlier years, 1983 moves slowly, suggesting a slowing-down of progress and a deceleration in the development of both Elvis and, by extension, Nigeria as a whole. The sense of time established by this formal structure might be described as a temporality of indebtedness, in which the past continually intrudes upon and makes demands of the present. Left unacknowledged, it threatens the present. This gradual grinding to a halt of personal and national progress is symbolically expressed by the laborious pace of life in Lagos itself. Early in the novel Elvis wakes up to 'the sound of babies crying, infants yelling for food and people hurrying but getting nowhere' (Abani 2004, p. 4). Held back by the city's inadequate infrastructure, which is epitomised by the city's famous 'go-slows' that immobilise traffic for hours, the residents of Lagos expend great energy for little returns. Their daily frustrations are symptomatic of larger systemic problems with Nigeria's development. Having borrowed vast sums from foreign lenders such as the World Bank, the leaders of the postcolonial state are perpetually looking over their shoulders, trying to retrospectively plug the gaping hole of national debt they have incurred. Fiscal irresponsibility coupled with a failure to acknowledge past mistakes results in the stagnated social situation that Abani describes. By formally arresting what Benedict Anderson in his well-known phrase described as the 'homogenous, empty time' of the nation (Anderson 1991, p. 26), he is able to demonstrate that Nigeria's decline was not inevitable, but the result of an incompetent state response to unwelcome neocolonial intrusions.

Abani's depiction of Nigeria's suspended modernisation resonates with anthropologist James Ferguson's recent analysis of African modernity in *Global Shadows*, in which he argues that the failure of many African societies to advance up the global political–economic hierarchy in the period following decolonisation – a failure that has multiple intrinsic and extrinsic causes – indicates that 'modernity' can no longer be understood as the teleological outcome of a supposedly natural process of national development, but rather as 'a standard of living, as a status' that is available to those who are already rich and powerful, but perpetually denied to the poor and marginalised (Ferguson 2006, p. 189). Especially pertinent is his assertion that:

> in a world of non-serialised political–economic statuses, the key questions are no longer temporal ones of societal becoming (development, modernisation), but spatialised ones of guarding the edges of a status group – hence the new prominence of walls, borders and processes of social exclusion in an era that likes to imagine itself as characterised by an ever expanding connection and communication.
>
> (Ferguson 2006, p. 192)

Through his characterisation of both Elvis and his mentor Redemption, Abani engages with both sets of questions. Resisting generic convention, Abani portrays Elvis not as an increasingly mature *Bildungsheld* (protagonist), but as an uncertain teenager who fails to live up to the standards of self-formation set by his community, his family and himself. If, for Esty, the 'frozen youth' of protagonists in such colonial *Bildungsroman* as *Kim*, *Lord Jim* and *An African Farm*, represents the necessarily stalled development of the periphery, Elvis' halting self-awareness similarly reflects the contradictory discourses of development by which he is interpellated (Esty 2007, p. 423). Whereas Elvis lacks the ability to read the city and interpret its many boundaries, Redemption navigates them with confidence, demonstrating a spatial awareness more appropriate to the suspended modernity that Abani depicts. Like Ferguson, Abani recognises that transnational cultural exchange can facilitate the crossing of apparently intransigent borders. However, as will be seen in my later discussion of the novel's conclusion, he has similar reservations about advancing a narrative of global mobility that obscures the mediating influence of restrictive social and economic conditions.

(Re-)imagining the margins

Nigeria's contradictory temporality of indebtedness is materialised in the urban structure of Lagos itself. In keeping with Abani's formal rendering of the deceleration and eventual stagnation of national progress through the differently paced narration of Elvis' Afikpo childhood and his time in Lagos, the structure of the city he describes is similarly uneven. Urban development is staggered according to the economic geography of the city, which Elvis perceives as 'half slum, half paradise' (Abani 2004, p. 7). On the one hand, he notes, 'Lagos did have its fair share of rich people and fancy neighbourhoods' characterised by 'beautiful brownstones in well-landscaped yards, sprawling Spanish-style haciendas in brilliant white and ocher, elegant Frank Lloyd Wright-styled buildings and foreign cars' (Abani 2004, p. 7–8). The residents of such properties can afford both the time to carefully plan the layout of their homes and the space in which to execute their elaborate designs. The carefully cultivated aesthetic of such houses provides a stark contrast with the cramped, haphazard construction of the shacks found throughout the city's slums, like Maroko and Aje, 'one of Lagos's oldest ghettos' that 'had no streets running through it, just a mess of narrow alleys that wound around squat, ugly bungalows and shacks' (Abani 2004, p. 51).

Although the slums' comparative disorder appears to be in total opposition to the city's wealthier neighbourhoods, Elvis quickly learns of their intimate connection. Sitting outside a local bar one evening, Elvis and Redemption gaze across the lagoon at the lights of Ikoyi, one of the most affluent parts of the city. Amused by the proximity of the slum-dwellers to their wealthy neighbours, Redemption asserts that this is why he likes Lagos: 'Because though dey hate us, de rich still have to look at us. Try as dey might, we won't go away' (Abani 2004, p. 137). Although Redemption is more than aware of the perils that his social marginalisation entails, his cynical humour correctly identifies the huge contradiction on which that exclusion rests. Although Lagos's rich might wilfully ignore the residents of slums such as Maroko, theirs is a precarious denial for they are massively indebted to their less fortunate urban neighbours. The luxuries of the city's wealthiest residents are made possible by the physical labour of urban immigrants like Elvis who accepts tenuous and poorly paid employment to make ends meet. On arrival in Lagos in 1983, his job prospects, like those of many rural to urban migrants, are particularly bleak. Although the oil trade had brought relative wealth to Nigeria in the 1970s, Elvis moves to Lagos during a period of severe economic depression following a sharp decline in oil prices and increased interest on Nigeria's IMF loans. With few jobs available, he seeks out casual work on a construction site: 'Lagos was littered with sites like this one, because new high-rise apartment complexes and office blocks were going up seemingly overnight' (Abani 2004, p. 27). Before long, Elvis loses his job on the pretext of his being a 'habitual latecomer' – ironic in the context of his own stunted development – but the nepotistically appointed site manager has actually been instructed by his father to 'lay off as many people as he could – something about being over budget' (Abani 2004, p. 72). Clearly, the construction of desirable living conditions for the rich is made possible by exploiting the livelihoods of the city's poor. However, this episode further demonstrates the tenuousness of the conspicuous wealth signalled by the new buildings. Reading the changing urban landscape as a symbol of Nigeria's social order, this poorly managed construction site suggests the dangers of sacrificing long-term plans for immediate profits.

When Elvis loses his construction site job, he finds himself in an economic predicament that is characteristic of what Mike Davis terms 'urbanisation without industrialisation' (Davis 2005, p. 14). Although he is relatively educated, there is little demand for Elvis' literacy skills. Like numerous other slum-dwellers, he has only his physical labour to offer, of which there is a surplus because Nigeria's industrial development has not kept pace with the urban growth of Lagos. In view of the accumulation of unemployed and casual workers in the slums of sprawling mega-cities such as Lagos, Davis observes that 'the principal function of the Third World urban edge remains as a human dump' (Davis 2005, p. 47). In a pertinent reading of *Graceland*, Ashley Dawson makes the related claim that 'Elvis and the other characters in Abani's novel constitute the violently evacuated waste products of today's world economy' (Dawson 2009, pp. 20–21).[12] The slum-dwellers' physical surroundings certainly seem to affirm their outcast status. As Elvis wakes up in his shack in the very first scene of the novel, he notes that 'the smell of garbage from refuse dumps, unflushed toilets and stale bodies was overwhelming' (Abani 2004, p. 4). Constructed on low-lying swampland that is dangerously prone to flooding, the entire slum occupies ambiguous territory. Elvis notes that he and his father are fortunate to 'live in one of the few places where Maroko made contact with the ground' (Abani 2004, p. 165). However, 'their short street soon ran into a plank walkway that meandered through the rest of the suspended city' (Abani 2004, p. 6). Despite the abundance of 'green swampy water that the ghetto was mostly built on' (Abani 2004, p. 4), the slum-dwellers have inadequate drinking water. Sanitation standards are predictably poor, exacerbated by the large number of people living in a small area.

Although, as Dawson argues, 'the shit in which they live is an apt metaphor for the social status of contemporary slum dwellers' (Dawson 2009, p. 20), Abani identifies some potential for resistance in their creatively improvised construction techniques. Headed to work one morning, Elvis, in a characteristically reflective moment, muses that,

> nothing really prepared you for Maroko. Half of the town was built of a confused mix of clapboard, wood, cement, and zinc sheets, raised above a swamp by means of stilts and wooden walkways. The other half, built on solid ground reclaimed from the sea, seemed to be clawing its way out of the primordial swamp, attempting to become something else.
>
> (Abani 2004, p. 48)

Clearly, the precarious living conditions of this so-called 'suspended city' materialise the social vulnerability of the Maroko residents (Abani 2004, pp. 6, 24). The unstable composite homes described here mark their literal absence of a foothold in the city. However, their tenacity in the face of desperate conditions is reflected in the aspirational quality of those shacks built on solid ground. Literally dragged down by their environment, the slum-dwellers seek and find means of survival.

Although the slum-dwellers demonstrate a resourceful ability to cope with their degraded physical surroundings, Abani suggests that their apparent reduction to human 'waste' is most pernicious in its disturbing psychological effects. Downtrodden by their economic, social and spatial marginalisation, some urban poor internalise their seeming lack of value, resulting in a nihilistic disregard for their own personal safety. Elvis encounters this fatalistic mentality while travelling home from work. During his journey, the bus he is on hits and kills a pedestrian. Such accidents are depressingly commonplace because numerous pedestrians ignore the many overhead bridges that cross the city's deadly motorways, leaving the 'the road littered with dead bodies at regular intervals' (Abani 2004, p. 57). Frustrated by his fellow Lagosians' willingness to concede their status as trash by putting themselves in mortal danger, he asks the old man sat next to him: 'Why do we gamble with our lives?' (Abani 2004, p. 57). The response he receives confirms the city residents' dogged resignation to their fate: 'We all have to die sometimes, you know. If it is your time, it is your time. You can be in your bed and die. If it is not your time, you can't die even if you cross de busiest road' (Abani 2004, p. 57). Although critics of urban culture such as Michel de Certeau have optimistically suggested that 'ordinary practitioners of the city' subvert imposed power structures by using spaces in ways that were unintended (de Certeau 1998, p. 93), what we see here is in fact a downtrodden compliance. Convinced of a dismal fate, the marginalised of Lagos are depressingly carefree. Ironically, Matthew Gandy suggests that the 'crumbling remains' of the city's 'concrete network of bridges, viaducts, flyovers, and cloverleafs . . . represent perhaps the most striking legacy of the oil boom' (Gandy 2005, p. 44). Initiated in the 1970s, such infrastructural programmes fell into disrepair during the early 1980s as a result of increasing national debt. While the bridges symbolise the arrested development of the nation as a whole, it is the urban poor who most directly confront Nigeria's economic decline. The state's absolute disregard for these people is further emphasised by the old man's reminder that the State Sanitation Department is too busy using its ambulances to fulfill private deals to collect the bodies from the road. The dead are left to rot like trash, while local authorities continue to profiteer in the face of social collapse.

Unlike his bemused travelling companion, Elvis vocally critiques his fellow Lagosians' apparent submission to their dire circumstances: 'That is the trouble with this country. Everything is accepted. No dial tones or telephones. No stamps in post offices. No electricity. No water. We just accept' (Abani 2004, p. 58). His frustration with the city's deep-rooted structural problems prompts the old man to ask 'Is dis your first day in Lagos?' (Abani 2004, p. 57), a question which indicates Elvis' naivety about the metropolis. However, his seeming inability to 'get up to speed' with how Lagos operates is also indicative of Elvis' struggle to come to terms with the city's arbitrary laws and obvious deficiencies. Unable to relinquish the idea that things could get better, Elvis holds on to the notion of possible progress that many of his urban peers, including Redemption, have abandoned. As Achille Mbembé explains, many Africans who find themselves in a 'situation of chronic scarcity' are compelled to relinquish long-term plans and gradual accumulation in favour of a 'course of life [that] is assimilated to a game of chance, a lottery, in which the existential temporal horizon is colonised by the immediate present and by prosaic short-term calculations' (Mbembé 2002, p. 271). Indeed, exhibiting a mentality closer to that of the pedestrians, Redemption is an avid gambler who regularly wins enough money to pay his rent. Elvis on the other hand always loses, underlining his inability to assimilate to the risk-taking mentality of the metropolis. Whereas Redemption takes lucky chances as meaningful signs – for example, when the two of them walk away from a motorcycle accident unscathed it is a 'good omen' (Abani 2004, p. 194) – Elvis seeks to impose meaning on the chaotic urban landscape which leaves little room for interpretation. Even at the very end of the novel, he thinks, 'there is a message to it somewhere, he mused, a point to the chaos. But no matter how hard he tried, the meaning always seemed to be out there beyond reach, mocking him' (Abani 2004, p. 307).

In his refusal to accept the status quo, Elvis exhibits the 'special sensitivity' characteristic of many classic *Bildungsroman* protagonists (Vázquez 2002, p. 86). Although an outsider, he sees Lagos with a clarity that many of its residents lack. Alluding to the subgenre of the *Künstlerroman* that narrates the formation of a central artist figure, Abani suggests that it is Elvis' own artistic inclinations that lend him his distinctive perspicacity. After his conversation with the old man on the bus:

Elvis could hardly wait for his stop and trudged home wearily, shoes ringing out on the walkways. It was late and much of Maroko was asleep, awash with moonlight. In the distance a young woman sang in a sorrow-cracked voice that made him catch his breath, stop and look around. In that moment, it all seemed so beautiful like a sequence from one of the films he had seen. Then the silence was broken by the approach of menacing steps. He turned and saw several figures running towards him.

(Abani 2004, p. 58)

The ominous interruption of this cinematic moment echoes an earlier episode in the novel during which Elvis' directorial musings on how best to frame the 'unflagging' energy of another shantytown are cut short by the attempted mugging of a young woman (Abani 2004, p. 29–30). By positioning Elvis as a spectator, Abani emphasises his detachment from his surroundings. His desire to direct the scenes before him speaks to a latent desire to control the city which he does not yet fully understand. By translating the chaos of Lagos into visual art, Elvis recasts reality in a manageable form. Yet in both instances, his artistic impulses are curtailed by the actual violence of street life. Here, Abani suggests the limitations of art as solace. On the one hand, the sympathetic perspective of his protagonist locates a poignant serenity in a slum that is painted in apocalyptic terms by the local government. Yet, if Abani himself is concerned with counteracting development discourse by presenting a more sensitive depiction of the urban margins, his self-reflexive characterisation of Elvis as budding artist suggests the dangers of detached aestheticisation. Indeed, as his protagonist is exposed to more extreme acts of urban violence, the capacity of aesthetic form to make sense of such atrocities is increasingly called into question.

Although Elvis fails to successfully integrate into the city, he continues to serve as a voice of conscience throughout the novel. His exchange with the old man on the bus is echoed later in the novel when he witnesses the disturbing administration of street justice to a suspected marketplace thief. Horrified by the thief's subjection to a fatal 'necklace of fire' in which a burning tyre is placed around his neck, Elvis looks to Redemption for an explanation of the crowd's unrelenting vengeance. In response to Redemption's assertion that their collective anger results from a lack of food and money, Elvis ponders, 'How long can we use the excuse of poverty?' (Abani 2004, p. 226). In a direct echo of the bus passenger's words, another unnamed witness to the thief's punishment interjects by asking Elvis, 'You dis man, you just come Lagos?' (Abani 2004, p. 226). Once again, Elvis' refusal to accept the dehumanising effects of poverty by conceding to an act that is 'purely animal' marks him as different, although he too shares the crowd's hunger and uncertainty (Abani 2004, p. 225).

Despite Elvis' concerned question, 'Nobody moved or spoke, not in the crowd, the buka or at the police checkpoint' (Abani 2004, p. 227). This collective immobility suggests the disturbing consensus that the challenging urban environment has produced. The police who 'watch the scene with bored expressions' represent the apathy of a state authority that has beaten its citizens into acceptance of their own worthlessness. They tolerate this act of ritual scapegoating because it redirects anger away from themselves. Yet as Elvis leaves the scene, his reaction suggests that the spectators' seeming apathy is symptomatic not of dull subjection, but collective trauma:

> As he climbed into the truck, Elvis was shaking. This scene had affected him more than anything else he had seen, though he wasn't sure why. Maybe it was the cumulative effect of all the horror he had witnessed; there was only so much a soul could take. As they drove off, Elvis watched the spreading fire through the tinted glass. It was horrifying, yet strangely beautiful.
>
> (Abani 2004, p. 228)

Elvis' physical reaction to the thief's death suggests how deeply disturbed he is by this incident. The fact that he is able to find beauty in the flames is therefore surprising. In a recent interview, Abani explains that 'a theme you'll find running through my work is the notion that the road to the sublime is through the grotesque' (Aycock 2009, p. 7). For Elvis, witnessing the 'necklace of fire' marks a significant step in his self-formation, as indicated by Abani's lengthy description of this brief scene at the very beginning of the novel's second book. If, according to Abani, *GraceLand* 'is a story about a loss of innocence, and yet that being some sort of redemption', this episode marks the beginning of Elvis' own loss of innocence (Aycock 2009, p. 8). When Elvis sees the flames as 'strangely beautiful', his reaction is not inhumane, but it is instead a paradoxical assertion of his humanity in spite of the horror to which he has been exposed.

Performing postcolonial urban identities

Although Elvis' critical urban gaze provides the reader with valuable insight into the routines, idiosyncrasies and dangers of Lagos, it sets him apart from the various social spheres in which he operates. Alienated from his family, his fellow slum-dwellers and Lagos at large, Elvis lives an unfulfilled half-life, unsure of his surroundings, but lacking an escape route. His existential uncertainty resonates with Wangari wa Nyatetu-Waigwa's 'liminal model' of the Francophone-African *Bildungsroman* (Wangari 1996). Drawing on cultural anthropologist Victor Turner's analysis of traditional African rites of passage, which he characterises as moving from social separation to marginalisation, culminating in incorporation, Wangari argues that the protagonists of African *Bildungsromane* 'remain suspended in the middle stage' (Wangari 1996, p. 3) which is characterised by 'symbolic invisibility or death' (p. 2). Abani underscores Elvis' marginalisation through the recurrent motif of spectrality. Elvis spends his time 'haunting markets and train stations, as invisible to the commuters or shoppers as a real ghost' (Abani 2004, p. 14). When he catches sight of his reflection in a puddle, his face 'seemed to belong to a stranger, floating there like a ghostly head' (Abani 2004, p. 6). Not only is Elvis estranged from others, but he is also unsure of himself. His detachment from his own reflection suggests a profound confusion about his own identity. The working out of such existential uncertainty is typically central to the *Bildungsroman*, as critic John Walsh explains:

> The picture [Bild] one has of oneself helps form an understanding of the world. The hero of the Bildungsroman builds an image of himself in his society and is ready to live in it, and this is how the novel ends.
>
> (Walsh 2008, p. 187)

In the same way that Elvis attempts to reframe the city around him in order to better understand it, he also tries to remould his own identity throughout the course of the novel. However, in contrast to the classic European *Bildungsroman*, his ability to project a satisfactory self-image is constrained by both local and global limitations on his self-realisation.

Elvis' desire for self-refashioning is evident in his dream of becoming a professional dancer. Lacking the confidence to audition for a proper dance troupe, he attempts to earn some money by impersonating his American namesake for the foreign tourists that visit Lagos. Although they do not appreciate his talent, offering him chocolate in return for his 'spellbinding' routine (Abani 2004, p. 12), Elvis finds comfort in putting on his costume and make-up. After a difficult day, his worries 'slip away' as he applies powder, mascara and lipstick in the privacy of his bedroom (Abani 2004, p. 77). Hidden from view, Elvis takes great pleasure in manipulating his appearance. Wearing make-up lends him some illusory control over the image (*Bild*) that he presents to the world.

The fact that Elvis wishes to make a living by impersonating a white American rock and roll singer might be interpreted as a symptom of Nigeria's neocolonial subjection to the import of American popular culture. The comfort Elvis derives from his make-up is short-lived since he can never truly resemble his namesake. Frustrated with the visible limitations of his meticulously applied maquillage, Elvis asks himself whether his life would have been any different 'if he had been born white, or even just American' (Abani 2004, p. 78). Without allowing himself to fully reflect on this question, he abruptly reminds himself that 'if Redemption knew about this, he would say Elvis was suffering from colonial mentality' (Abani 2004, p. 78). In his desire to mimic the famous looks and dance moves of Elvis Presley, Abani's Elvis seems to pander to a widespread desire for the replication of American culture in Nigeria. Indeed, the entire city in which he lives is apparently in thrall to the fashion, architecture and cars of the United States, as the reader learns early in the novel: 'Name it and Lagos had a copy of it, earning it the nickname "One Copy"' (Abani 2004, p. 8).

However, as with any cultural export, the arrival of American popular culture in Africa is a mediated process. Elvis' interest in dancing is, in part, a tribute to his mother who chose his name because she was a Presley fan. Furthermore, Elvis is inspired to become a professional Elvis impersonator when, as a young boy, he witnesses some Ajasco dancers performing to Presley's 'Hound Dog' in the Afikpo market. It is their striking outfits – 'high wigs, dark sunglasses, and white long-sleeve shirts, gloves, trousers and white canvas shoes' – and 'fluid' dance moves that entrance Elvis, persuading him of 'what he wanted to do more than anything else' (Abani 2004, p. 65). The Elvis that he wishes to impersonate is a vision that has been refracted, indigenised and transformed during its importation to Nigeria. As Jacob Patterson-Stein (2009) suggests, Abani uses Elvis' fixation with an icon of American music to demonstrate the untenability of static national cultures. American music, he argues, following Homi Bhabha, creates a de-nationalised 'third space' in the novel, which suggests the possibility of forging more flexible, transnational cultural identities. In this view, Nigeria is not the passive recipient of a more 'developed' American culture, but rather an agent in a process of mutually transformative cultural negotiation.

In addition to challenging the fixed locations of national cultures, Elvis' dance routine provides him with a significant opportunity to experiment with the performance of his own gender identity. In doing so, he refuses the social conformity that the classical *Bildungsroman* advocates. In Moretti's understanding of the genre as an inherently conservative 'symbolic form' in which the benefits of bourgeois socialisation outweigh those of unrestricted individuality, the protagonist must learn to live with 'the *interiorisation of contradiction*. The next step being not to 'solve' contradiction, but rather to learn to live with it, and even transform it into a tool for survival' (Moretti 2000, p. 10). In contrast, by displaying a range of masculine and feminine attributes through costume and performance, Elvis makes visible his own fluid gender identity.

By disrupting conventional gender codes, Elvis highlights the dangerous reactionism of men like his father who turn to prohibitively conservative cultural models in response to the social changes produced by neocolonial modernity. As a child, Elvis takes part in a confusing male initiation ceremony, intended to mark 'his first step to manhood as dictated by tradition' (Abani 2004, p. 18). Too young to comprehend why his father and uncle 'strapped a grass skirt on him and then began to paint strange designs in red and white dye all over his body' (Abani 2004, p. 17), Elvis' later make-up routine is foreshadowed when the elders 'blew chalk powder in his face' (Abani 2004, p. 20). Although Sunday later expresses outrage at Elvis' apparent betrayal of his essential masculinity through his supposedly effeminate dancing, the element of costuming and make-up central to this earlier ceremony affirms the inherent contingency of gender identities, which always emerge from complex social performances rather than inherent physical or psychological traits.

Instead of killing an eagle during the ceremony as his ancestors did, his father hands Elvis an already-pierced chick that he has strung onto a homemade bow. As his uncle Joseph explains, 'eagle is too expensive' (Abani 2004, p. 19). Although Sunday informs him that 'dis is about being a man,' the nervous Elvis is hesitant to accept this poor substitution (Abani 2004, p. 18). In her reading of *GraceLand* as a 'trauma novel', Amy Novak identifies this scene as an example of how 'the erasure of traditional culture by colonialism is followed by the inroads of Western culture in post-Independence/neocolonial Nigeria' (2008, p. 37). It is not only Elvis who is confused and scared; more interested in consoling himself with whiskey than Elvis' wellbeing, Sunday is similarly struggling to come to terms with the cultural estrangement wrought by colonial occupation and its enduring effects. Daunted by the uncertain future for which the ceremony ostensibly prepares his son, Sunday thus resorts to an essentialist notion of masculinity in an ultimately damaging attempt to delineate a stable social role for himself and Elvis.

Abani furthers his critique of Sunday's cultural conservatism through the inclusion of textual fragments which preface each chapter. Apparently taken from an ethnographical work, these brief excerpts first describe the Igbo kola nut ritual from an insider's perspective then analyse it from the perspective of a sympathetic outsider. Performed on many occasions, the ceremonial breaking of the kola nut welcomes visitors, blesses gatherings and seals covenants. The ostensible effect of these short paragraphs is to celebrate the richness of Igbo culture in all its complexity. Cultural commemoration is inherent to the ritual itself, which is 'part hospitality, part etiquette, part protocol and part history lesson' (Abani 2004, p. 172). As the ethnographic narrator explains: 'Every time the ritual takes place, the history of all the clans present, and their connections, is enacted. This helps remembering' (Abani 2004, p. 240). Unlike the temporality of indebtedness to which national history and the postcolonial present are tethered, this ritual provides a sense of comfort and stability by marking social continuity in a recognisable, cyclical fashion.

Although broadly sympathetic to Igbo culture, Abani's inclusion of these excerpts resists the simplistic idealisation of Igbo tradition. In a recent interview, he explains that the 'irritating voice' of the ethnographer captures the cultural simplification of anthropologists who 'believe that if you can figure out this one ritual, you can understand the Igbo. . . [I]f you figure out one aspect of this complicated place, then that gives you the key to everything' (Aycock 2009, p. 8). By ironically juxtaposing these excerpts alongside Elvis' undignified Lagos existence and his failed coming of age, Abani quietly asserts a further critique. Not only does he highlight the contemporary degradation of Igbo traditions, but he also suggests flaws in the gender divide on which they are based. The kola nut ceremony is exclusively male, as the ethnographer notes: 'Women take no part in the kola nut ritual. In fact, female guests are never presented with kola nuts' (Abani 2004, p. 172). Although, according to the ethnographer, 'the kola-nut ritual provides a ritual space for the affirmation of brotherhood and mutual harmony' (Abani 2004, p. 208), Elvis' contrasting experience of the initiation ceremony has the opposite effect, alienating him from his male relatives and peers.

'Ritual space' is even more compromised in 1980s Lagos than it is in the Afikpo of Elvis' early childhood. Indeed, it is almost entirely absent. The cultural coordinates provided by such rituals as the kola nut ceremony have largely disappeared in Maroko, where Sunday's veranda becomes the site of 'the routine most evenings' with various neighbours gathering there to drink alcohol after nightfall (Abani 2004, p. 80). Abani punctures the frequently made distinction between supposedly modern urban lifestyles and a more traditional rural way of life by highlighting the tenacious marginalisation of women that persists even in Maroko:

> The women sat like shadows behind the men and seemed to use the fact that they needed the light to darn, or shell melon seeds, to justify their presence. For the most part, the men ignored them. Those brave enough to call their husbands' attention to something were rewarded with a gruff and impatient answer, as though they were keeping the men from some important philosophical breakthrough.
>
> (Abani 2004, p. 80)

The Maroko residents' recourse to these unequal, but familiar gender roles suggests a desire to preserve a minimal amount of social control despite their changed circumstances. The nightly gatherings on Sunday's veranda lend the Maroko residents some social cohesion, albeit in a problematic form. Whereas the kola nut ceremony facilitated deep reflection, this nightly 'palm wine ceremony' promotes little more than loud arguments and casual gossip. With the contraction of lived space in Maroko comes the dilution of collective rituals. Even if an argument might be made in favour of gender division upholding social harmony in the past, the Maroko residents exhibit a wilful reluctance to accept the inherent flexibility of Igbo tradition, which, as the ethnographer notes, 'is fluid, growing. It is an event, like the sunset or rain, changing with every occurrence' (Abani 2004, p. 291). Unable or unwilling to adapt to their new urban environment, the residents of Maroko insist on outdated gender roles. These are as inimical as development discourse in their recuperation of false traditions. This is especially problematic for Elvis, who is held to a redundant model of masculinity that inhibits his pursuit of the professional dancing career he wants.

Grassroots resistance, global struggle

Although Elvis' performative desires allow Abani to hint at the constructedness of cultural and gender identity, the economic pressures to which Elvis is subjected as a slum-dweller curtail their liberating potential. Chastised by his father and dismissed by the tourists who are his potential audience, Elvis realises that his dance routine cannot secure him the regular income necessary to fulfill his basic needs in Maroko. Having decided to look for a steady job, Elvis reflects on the temporary postponement of his artistic aspirations:

> It wasn't that he was giving up on his dream to be a dancer, he rationalised; it was more like he was deferring it for a while. Maybe with the money he earned he could save up to go to America. That was a place where they appreciated dancers.
>
> (Abani 2004, p. 24–25)

Elvis' attempted self-reassurance invokes Langston Hughes's 1951 book-length poem *Montage of a Dream Deferred*, a polyphonic expression of the frustrations and aspirations of a marginalised black urban population in segregated postwar America. By implying a comparison between Elvis' own arrested development and that of mid-twentieth-century African Americans, Abani cultivates a transatlantic and transhistorical network of diasporic cultural resistance that is at once liberating and limited. While African artistic solidarity with black American writers is potentially inspirational and motivating, the fact that Elvis remains trapped in similarly oppressive circumstances in 1983 indicates the extent of the global struggle for equal human rights still to be fought. His dream of emigrating to America is ironised by the long history of stateside inequality that Abani alerts us to, in which the economic and social discrimination to which Elvis is accustomed intersects with racism. By suggesting the common historic legacy of marginalisation to which both the residents of Harlem and those of Maroko are subject, Abani invites the reader to see Elvis' struggles through a global lens. Unlike development advocates who posit African 'exceptionalism' as the cause of the continent's economic decline, Abani suggests that the perceived 'backwardness' of such cities as Lagos does not result from local cultural inclinations, but that it is instead symptomatic of a global system of oppression that casts the urban poor aside.

Hughes anticipates Abani's challenge to the suspended temporality of urban marginalisation, notably in the 'Harlem' section of his poem:

> What happens to a dream deferred?
> Does it dry up
> like a raisin in the sun?
> Or fester like a sore –
> And then run?
> Does it stink like rotten meat?
> Or crust and sugar over –

like a syrupy sweet?
Maybe it just sags
like a heavy load.
Or does it explode?

(Hughes 1990, p. 268)

The subtle defiance of Hughes's initial question is gradually reinforced by the subsequent inquiries that evoke images of decay, pollution and death. The final lingering inquiry, 'Or does it explode?' is particularly disturbing in its understated threat of violence. The cumulative effect of the questions posed in 'Harlem' is that of the speaker's restrained aggression, held back by the disappointment evoked in the opening line. The final section of *GraceLand* offers several responses to these questions in its depiction of the different fates of Elvis and his father as they both encounter the crippling intransigence of the state and its willingness to exert violent force in order to suppress dissenters.

Sunday's role in the novel culminates in his ultimately suicidal protest against the government's demolition of Maroko. The destruction of the slum is an undeniable act of violence that is, following Ferguson, designed to 'guard the edges of a status group' (Ferguson 2006, p. 192); in this case, both the physical and figurative boundaries of the state authorities and the urban elite. Refusing to accept this blatant disavowal of his own status as an urban resident, Sunday is finally spurred into action. However, the fatal consequences of his protest suggest that more than a sudden 'explosion' is needed to preserve dreams – long-term planning is necessary.

The terms under which Maroko is slated for destruction demonstrate the postcolonial state's extension of imperial development discourse. Rasaki's government pathologises the slum in the manner of European colonisers concerned with maintaining boundaries between putatively dirty and dangerous 'native' areas of the city and those in which whites resided. Reading the newspaper one morning, Sunday notices that what he refers to as 'dis crazy government' has plans to flatten the slum. As he explains, 'dey say we are a pus-ridden eyesore on de face of de nation's capital' (Abani 2004, p. 247). He is immediately alert to the discursive artillery that the government intend to mobilise alongside their bulldozers, observing that 'Dey even have a military sounding name for it – Operation Clean de Nation' (Abani 2004, p. 247). Despite the government's attempt to pass off the destruction of the slum as an effort to ensure public health, Sunday recognises the sanitation initiative as an attack on his home and neighbours. The martial character of the government's development initiative echoes the observation of Mama Put, a character in Wole Soyinka's play *The Beatification of Area Boy* (1995), which also portrays the forced evacuation of Maroko. A survivor of the Biafran War who has moved to Lagos, Mama Put observes that the demolition of Maroko is 'war of a different kind. It is war of a kind governments declare against their people for no reason' (Soyinka 1995, p. 76). By inviting a similar comparison between the forced removal of thousands of slum-dwellers and the displacement of wartime refugees, Abani characterises the demolition of Maroko as a form of senseless civil conflict which further belies the unity of the Nigerian nation. The vision that Lagos's military government holds for a clean, modern city cannot be materialised without multiple acts of violence. Implementation of its plans mandates not only crushing the slum-dwellers' grassroots resistance to their removal, but also the discursive erasure of Maroko's long history. In seeking to implement what Michel de Certeau has described as the 'Concept-city' produced by city planners – a two-dimensional urban map that figuratively flattens the space of the city – they must literally eradicate all traces of the slum's palimpsestic urban history (de Certeau 1998, p. 95). In his detailed account of the Maroko residents' opposition to the slum's demolition, Abani adds to the critique of this developmental amnesia initiated by the structure of his novel.

Sunday emerges as the accidental leader of a grassroots campaign to prevent the demolition of Maroko. The veranda on which he spent many hours drowning his sorrows becomes the site of revolutionary strategising, much to Elvis' bemusement (Abani 2004, p. 253). Although latent gender discrimination persists in his allocation of different roles to male and female protesters, Sunday succeeds in marshalling significant community support by using the canvassing skills that earlier failed to ignite his election campaign in Afikpo. The subordination of his self-interest to the protection of the entire community suggests one possible reason for his positive reception. Although the state authorities are far better equipped than the slum-dwellers with a raft of bulldozers on hand to rapidly crush the shacks, the Maroko residents succeed in sabotaging their initial demolition attempts. Drawing a parallel between the Maroko demonstrations and contemporaneous 'IMF riots' that took place during the 1980s, Ashley Dawson suggests that 'this strand of *Graceland* highlights the power of spontaneous popular organisation and protest' (Dawson 2009, p. 30). However, as Dawson notes, it is a power that is quickly exhausted by unrelenting state violence. Indeed, Abani implies that the spontaneous resistant energies catalysed by this moment of crisis need to be harnessed for more sustained collective organisation. The manner in which he narrates Sunday's death suggests that doing so requires a response to not only developmental amnesia, but also the cultural forgetfulness of the slum-dwellers themselves. Having refused to move from his veranda, he is shot by a policeman and subsequently crushed by the bulldozer that destroys his home. However, at the moment of his death, Sunday assumes the form of the leopard totem that he thinks he has been hallucinating: 'Sunday roared, leapt out of his body and charged at the back of the policeman, his paw delivering a fatal blow to the back of the policeman's head' (Abani 2004, p. 287). The fact that Elvis later discovers the policeman's body covered in claw marks even though 'there were no animals of that size anywhere near Lagos or Maroko' invites the reader to take this inexplicable moment seriously. Gwendolyn Etter-Lewis has argued that Sunday's death implies the decline of the 'older, traditional African masculinities' that his totem represents (Etter-Lewis 2010, p. 171). However, this unique moment of magical realism in the novel implies a necessary reconnection with ancestral wisdom. As his depiction of Igbo gender roles suggests, Abani does not idealise traditional cultural knowledge, but nor does he wholly dismiss it. He suggests that there is a place for the ancient and ineffable in a modernising global society.

Elvis is absent from Lagos during the slum clearance, but on his return he participates in a protest performance in Tinubu Square, intended as a peaceful, but powerful call for democracy. Ironically the site of Nigeria's Amalgamation Ceremony in which the North and South protectorates were united to form a single nation in 1914, the square 'erupts' into civil violence when soldiers arrive to disperse the crowd, recalling Hughes's poetic premonition of explosive social unrest (Abani 2004, p. 287). Caught unawares, Elvis is 'completely confused' by what is happening, again demonstrating his naivety about the extreme means the government is willing to deploy in order to maintain control. This naivety is compounded when he is accosted by a soldier and ordered to identify himself. Only able to stutter 'I... I...' in response, he is taken into custody (Abani 2004, p. 288). His uncertainty about who he is and where he is proves near fatal as he taken for questioning at Tango City, the Special Military Interrogation Unit.

Elvis' imprisonment provides the novel's most explicit dramatisation of the manner in which the postcolonial state not only inhibits, but also actively prevents the development of Nigeria's marginalised urban poor. Held for days without charge, Elvis is relentlessly tortured by the military police who want him to reveal the whereabouts of his friend the Beggar King who they believe to be the protest ringleader. He is thrown into a cell where he is 'hung from the metal bars on the window, feet dangling six inches from the floor, suspended by handcuffs' (Abani 2004, p. 288). His literal suspension at the hands of the military police renders the state's inimical effect on his personal development explicit. Abani details their particular callousness as they use various techniques, such as smearing him with industrial disinfectant, to ensure he experiences the maximum possible pain when they beat him.[13] Prior to his arrest, the reader has witnessed Elvis' struggle to comprehend his urban surroundings and the postcolonial power dynamics that intersect the city. In prison, his quest for understanding is reduced to actual unconsciousness through flogging and torture. His 'tormentors' further work to undermine Elvis' tentative *Bildung* by challenging his maturity. They refer to him as a 'stupid boy' who is 'young and confused' (Abani 2004, p. 296), suggesting his lack of formation. They further challenge his masculinity by subjecting him to a perverse male initiation ceremony which they claim to have adopted from Fulani tradition in which young men are whipped 'to test who be man enough to marry' (Abani 2004, p. 295). Those who do not make a sound are considered to have reached adulthood. Although they beat Elvis to within an inch of his life, he cannot provide the information they want. The Colonel concludes that 'dis one is just a child' who can be 'thrown back,' like trash, onto the streets (Abani 2004, p. 296).

After his release from prison, Elvis returns to Maroko to find his home destroyed and his father dead (Abani 2004, p. 305). He eventually makes his way to another slum that has suddenly increased in size owing to the influx of refugees from Maroko. For the 'food sellers, soft-drink hawkers, tire vulcanizers, small-time car mechanics, women and men lying on top of their belongings and hundreds of beggar children' that crowd Bridge City, Operation Clean the Nation has only served to further tighten their living conditions (Abani 2004, p. 306). The slum clearance programme not only displaces the Maroko residents, but also those who have to make way for them in the other ghettoes around the city, suggesting the limited scope of the urban development it achieved.

In keeping with his role as Elvis' mentor, it is Redemption who provides Elvis with a way out of the depravity in Bridge City where young children are frequently 'beaten, raped, robbed, and sometimes killed' by giving him a forged US passport that he once intended to use for his own escape to America (Abani 2004, p. 309). However, Elvis' imminent departure from Lagos is far from an uplifting conclusion to the novel. Although this final scene suggests that he is about to achieve some liberation from the pressures of Lagos, he is doing so in the guise of someone else. Elvis' own identity remains insecure, contingent and concealed. While relocating from the Third to the First world suggests a forward trajectory up the global hierarchy, Elvis' reluctant emigration uneasily evokes the painful transatlantic ties established by the Middle Passage, a traumatic history that is explicitly invoked by Elvis himself who identifies with the character of a lynched black American in James Baldwin's *Going to Meet the Man*, which he is reading as he waits to board his plane. Adding another layer to his already troubled sense of self, Elvis' identification with 'a dying black man engulfed by flames' suggests the intransigence of the social inequalities that he will still have to navigate despite his departure from Lagos (Abani 2004, p. 319). Elvis' reluctance to leave stems from an unwillingness to relinquish what he describes as 'something essential' that he cannot explicitly describe (Abani 2004, p. 319). This ambiguous, but deep sensation recalls his response to the pain he experienced during his torture at the hands of the state police, the intensity of which made him feel like 'It was part of him now. It seemed like he couldn't remember a time when it was not here. It had become essential to him' (Abani 2004, p. 294). In his anticipation of further trauma in America, as symbolised by Baldwin's lynching victim, Elvis reaches a tentative understanding of his place in the world. Horrified by the genital mutilation of the man in the short story, Elvis sees himself as 'that scar, carved by hate and smallness and fear onto the world's face' (Abani 2004, p. 320). Like the thief who was subjected to the 'necklace of fire' by a 'mob of lynchers' (Abani 2004, p. 228), Elvis is figured as a scapegoat. Cast out of Lagos, he signifies the failure of the local, national and international discourses of development that intersect postcolonial Nigeria. When the airline clerk calls Elvis' new name, he responds in the final sentence of the novel with 'Yes, this is Redemption' (Abani 2004, p. 321). His reflection on Baldwin's harrowing story ironises the promise of this new identity by indicating the precarious salvation from persecution that America offers to young black men. Although Elvis knowingly considers that 'nothing is ever resolved, . . . it just changes', the ambiguous conclusion to Abani's novel avoids unsatisfactory quietism in its call for more diligent historical awareness (Abani 2004, p. 320). In her reading of the open-endedness of many third-generation Nigerian novels, Madhu Krishnan argues that 'flouting the traditional conception of closure allows these narratives to block any simple resolution to the past and its traumas and instead forces a lasting engagement with history and its effects' (Krishnan 2010, p. 186). This claim is further strengthened by considering Abani's particular disruption of the *Bildungsroman*'s expected resolution. Elvis' uncertain departure from Lagos defies what Apollo Amoko terms the 'retrospective logic' of the African *Bildungsroman* (Amoko 2009, p. 206). As he explains in a comparative analysis of African autobiography and *Bildungsroman*,

> Both genres seem, at some level, to begin at the end. . . . Notwithstanding any number of plot twists and turns, the Bildungsroman invariably seems to require, in the end, the protagonist's formation (or Bildung). The fact of eventual, if not inevitable, Bildung becomes anterior to, if not determinative of, the innumerable twists and turns that constitute the rest of the narrative.
>
> (Amoko 2009, p. 196)

As previously discussed, Elvis fails to mature according to generic convention. His rather hesitant *Bildung* takes the form of a gradual loss of innocence as he is exposed to the violent dynamics of contemporary postcolonial urban life. Reflected in the liminal shacks of the Maroko slum, his suspended development is analogous to that of the nation itself, which, as indicated by Abani's narrative structure, stalls and declines as a result of incompetent postcolonial government. By refusing narrative closure at the end of *GraceLand*, Abani rejects a conciliatory retrospective moment that would offer a benevolent resolution to the trials Elvis has experienced. Instead, he asserts the multidirectional histories of marginalisation that have been obscured by linear trajectories of urban, national and international development. Leaving the reader to speculate on Elvis' uncertain future, Abani calls for new measures of progress that do not selectively ignore the enduring wreckage of the past, but rather engage with, learn from and cautiously come to terms with it.

Notes

1 For a concise history of Lagos, see Olaniyan 2004, pp. 88–89. For extended treatment, see also Mabogunje (1968), Ajetunmobi (2003), Fasinro (2004), and Ajunam (2004).
2 Catherine Kehinde George (2010) lists the 'challenges arising from the rapid urbanisation of Lagos metropolis' as 'urban sprawl, encroachment on conservation zones, inadequate basic infrastructure and communal facilities, inadequate energy (electricity), inadequate potable water, formation of slums, urban road transport problems, urban traffic congestion, municipal waste management, urban violence, change of use/illegal development, and impact of industries'.
3 This 'exceptionalism hypothesis' is further explained in Martine *et al.* (2008, pp. 308–9).
4 For critical responses to Western development discourse, see Amin (1990), Sachs (1992), Escobar (1994).
5 See Soyinka's *Beatification of Area Boy* (1995), J. P. Clark Bekederemo's 'Maroko' (1999), Ofeimun's 'Demolition Day' (2000), and Nwosu's *Invisible Chapters* (2001).
6 Odia Ofeimun first identifies this 'Maroko Corpus' as such in his review of Maik Nwosu's *Invisible Chapters* (Ofeimun 2005).
7 See Agbola and Jinadu for a comprehensive list of Lagos evictions from 1973–95, including three separate clearances of Maroko (Agbola and Jinadu 1997, pp. 274–75).

8 As both the discourse and practice of development come under increasing scrutiny, scholars engaged in critical development studies have acknowledged 'that fictional accounts of development can sometimes reveal different sides to the experience of development and may sometimes even do a "better" job of conveying the complexities of development than research-based accounts' (Lewis *et al.* 2005, p. 6). Although one might caution against mining literary sources for objective 'data', a fuller consideration of the descriptive power of literary fiction responds to the need for analyses of development that extend beyond the measurement of economic criteria. Nigerian literary critic Molara Ogundipe-Leslie notes that 'development as conceived historically now in Western hegemonic discourses and activities is failing because it has no cultural face' (Ogundipe-Leslie 1998, p. 27). As she explains, development agendas, both internal and external, frequently ignore the beliefs, traditions, and principles respected by those whom they will most directly affect. Literary representations of development such as Abani's are uniquely placed to engage with the 'creation, questioning, rejection, and restructuring' of such values (Ogundipe-Leslie 1998, p. 28).

9 See Ngũgĩ's *Weep Not Child* (1964) and Dangarembga's *Nervous Conditions* (1988).

10 See, for example, Helon Habila's *Waiting for an Angel* (2002).

11 Of course, this is not to suggest that Shehu Shagari's presidency of the Second Republic (1979–83) was squeaky clean. See Falola and Heaton (2008, pp. 181–208) for a detailed historical overview of this period in Nigerian history, which suggests that the ruling National Party of Nigeria was, in fact, characterised by corruption.

12 For an extended sociological analysis of the 'wasted lives' created by global modernisation, see Bauman (2004).

13 Abani no doubt draws on his own experiences as a teenage political prisoner in Kiri Kiri maximum security prison here. As he explains in the author's introduction to *Kalakuta Republic* (Abani 2000) – his harrowing collection of poems dealing with his confinement and torture – he was imprisoned three times between 1985 and 1991, having been arrested for writing putatively treasonous literature (pp. 9–10).

References

Abani, C. (2000) *Kalakuta Republic*. London: Saqi Books.

Abani, C. (2004) *GraceLand*. New York: Picador.

Achebe, C. (1960) *No Longer at Ease* London: Heinemann.

Agbola, T. and Jinadu, A. M. (1997) 'Forced Eviction and Forced Relocation in Nigeria: The Experience of Those Evicted from Maroko in 1990', *Environment and Urbanization* vol. 9, no. 2, pp. 271–88.

Ajetunmobi, R O. (2003) *The Evolution and Development of Lagos State*. Lagos, Nigeria: A-Triad Associates.

Ajunam, A. (2004) *Eko: The Navel of the Giant*. Durban: LexisNexis/Butterworths.

Amin, S. (1990) *Maldevelopment: Anatomy of a Global Failure*. London: Zed Books.

Amoko, A. (2009) 'Autobiography and *Bildungsroman* in African Literature', in Irele, A. F. (ed.) *The Cambridge Companion to the African Novel*, Cambridge: Cambridge UP, pp. 195–208.

Anderson, B. (1991) *Imagined Communities: Reflections on the Origin and Spread of Nationalism*. London: Verso.

Aycock, A. (2009) 'An Interview with Chris Abani', *Safundi: The Journal of South African and American Studies*, vol. 10, no.1, pp. 1–10.

Bakhtin, M. M. (1986) 'The *Bildungsroman* and Its Significance in the History of Realism (Toward a Historical Typology of the Novel)', in *Speech Genres and Other Late Essays* (trans. V. W. McGee), Austin, TX: U of Texas P, pp. 10–59.

Bauman, Z. (2004) *Wasted Lives: Modernity and its Outcasts*. Cambridge: Polity Press.
Buckley, J. H. (1974) *Season of Youth: The Bildungsroman from Dickens to Golding*. Cambridge, MA: Harvard UP.
de Certeau, M. (1998) *The Practice of Everyday Life* (trans. S. Rendall), Berkeley, CA: University of California Press, 1998.
Clark Bekederemo, J. P. (1999) 'Maroko', in *A Lot from Paradise*. Lagos, Nigeria: Malthouse Press, pp. 44–45.
Crush, J. (1995) *Power of Development*. London: Routledge.
Dangarembga, T. (1988) *Nervous Conditions*. Banbury: Ayebia Clarke.
Davis, M. (2005) *Planet of Slums*. London: Verso.
Dawson, A. (2009) 'Surplus City: Structural Adjustment, Self-fashioning, and Urban Insurrection in Chris Abani's *Graceland*', *Interventions*, vol. 11, no.1, pp. 16–34.
Ekwensi, C. (1954) *People of the City*. London: Andrew Dakers.
Escobar, A. (1994) *Encountering Development: The Making and Unmaking of the Third World*. Princeton, NJ: Princeton UP.
Esty, J. D. (2007) 'The Colonial Bildungsroman: *The Story of an African Farm* and the Ghost of Goethe', *Victorian Studies*, vol. 49, no. 3, pp. 407–30.
Etter-Lewis, G. (2010) 'Dark Bodies/White Masks: African Masculinities and Visual Culture in Graceland, The Joys of Motherhood and Things Fall Apart', in Mugambi, H. and Allan, T. (eds.), *Masculinities in African Literary and Cultural Texts*. Banbury: Ayebia. pp. 160–177.
Falola, T. and Heaton, M. (2008) *A History of Nigeria*. Cambridge: Cambridge University Press.
Fasinro, H A. B. (2004) *Political and Cultural Perspectives of Lagos*. Lagos, Nigeria: Academy Press.
Ferguson, J. (2006) *Global Shadows: Africa in the Neoliberal World Order*. Durham, NC: Duke UP.
Gandy, M. (2005) 'Learning from Lagos', *New Left Review*, no. 33: pp. 37–52.
Habila, H. (2002) *Waiting for an Angel*. New York: W. W. Norton & Co.
Hughes, L. (1990) *Selected Poems of Langston Hughes*. New York: Vintage Books.
Kehinde George, C. (2010) 'Nigeria: Challenges of Lagos as a Mega-City', *Daily Independent*, 21 February [Online]. Available at http://allafrica.com/stories/201002221420.html (Accessed 30 December 2015).
Krishnan, M. (2010) 'Biafra and the Aesthetics of Closure in the Third Generation Nigerian Novel', *Rupkatha Journal on Interdisciplinary Studies in Humanities*, vol. 2, no.2, pp. 185–95.
Lewis, D., Rodgers, D. and Woodcock, M. (2005) 'The Fiction of Development: Knowledge, Authority and Representation', Working Paper 05-61, London: London School of Economics and Political Science [Online]. Available at http://eprints.lse.ac.uk/379/ (Accessed 3 January 2016).
Lukács, G. (1971 [1920]) *The Theory of the Novel* (trans. Anna Bostock), Cambridge, MA: MIT Press.
Mabogunje, A. (1968) *Urbanization in Nigeria*. London: U of London P.
Martine, G., McGranahan, G., Montgomery, M. and Fernández-Castilla, R., eds. (2008) *The New Global Frontier: Urbanization, Poverty and Environment in the 21st Century*. London: Earthscan.
Mbembé, A. (2002) 'African Modes of Self-Writing', *Public Culture,* vol.14, no.1, pp. 239–73.
Moretti, F. (2000 [1987]) *The Way of the World: The Bildungsroman in European Culture*. London: Verso.

Murray, M. and Myers, G. (2006) *Cities in Contemporary Africa*. London: Palgrave Macmillan.

Ngũgĩ wa Thiong'o. (1964) *Weep Not Child*. London: Heinemann.

Novak, A. (2008) 'Who Speaks? Who Listens?: The Problem of Address in Two Nigerian Trauma Novels', *Studies in the Novel*, vol. 40, no. 1, pp. 31–51.

Nwosu, M. (2001) *Invisible Chapters*. Lagos, Nigeria: House of Malaika & Hybun.

Ofeimun, O. (2001) 'Imagination and the City', *Glendora Review: African Quarterly on the Arts*, vol 3, no. 2, pp. 11–15, 137–41.

Ofeimun, O. (2000) 'Demolition Day', in *London Letter and other Poems*. Lagos, Nigeria: Hornbill House, p. 6.

Ofeimun, O. (2005) 'Daring Visions: *Invisible Chapters* by Maik Nwosu', *English in Africa*, vol. 32, no.1, pp. 135–41.

Ogundipe-Leslie, M. (1998) 'Literature and Development: Writing and Audience in Africa', in Adams, A. & Mayes, J. (eds.) *Mapping Intersections: African Literature and Africa's Development*, Trenton, NJ: Africa World Press, pp. 27–36.

Olaniyan, T. (2004) *Arrest the Music!: Fela and His Rebel Art and Politics*. Bloomington, IL: Indiana University Press.

Patterson-Stein, J. (2009) 'De-nationalizing American Music in the "Third Space" of *GraceLand*', *eSharp*, no. 13, pp. 48–68 [online]. Available at http://www.gla.ac.uk/media/media_122691_en.pdf (Accessed 3 January 2016).

Redfield, M. (1996) *Phantom Formations: Aesthetic Ideology and the Bildungsroman*. Ithaca, NY: Cornell University Press.

Rasaki, R. (1988) *Managing Metropolitan Lagos* [Online], Ota, Nigeria, Africa Leadership Forum. Available at http://africaleadership.org/rc/Managing%20Metropolitan%20Lagos.pdf (Accessed 30 December 2015).

Sachs, W. (1992) *The Development Dictionary: A Guide to Knowledge As Power*. London: Zed Books.

Soyinka, W. (1995) *The Beatification of Area Boy: A Lagosian Kaleidoscope*. London: Methuen Drama.

Slaughter, J. (2007) *Human Rights, Inc.: The World Novel, Narrative Form, and International Law*. New York: Fordham UP.

Vázquez, J. S. F. (2002) 'Recharting the Geography of Genre: Ben Okri's *The Famished Road* as a Postcolonial *Bildungsroman*', *The Journal of Commonwealth Literature*, vol. 37, no. 2, pp. 85–107.

Walsh, J. (2008) 'Coming of Age with an AK-47: Ahmadou Kourouma's *Allah n'est pas obligé*', *Research in African Literatures*, vol. 39, no.1, pp. 185–97.

Wangari wa Nyatetu-Waigwa (1996) *The Liminal Novel: Studies in the Francophone-African Novel as Bildungsroman*. New York: Peter Lang.

3 'A new heightened sense of place'

Dinaw Mengestu's cognitive map of Washington, D.C.

This chapter examines a novel that, in many ways, picks up the narrative thread that Chris Abani leaves poignantly hanging at the end of *GraceLand*. The latter concludes with the image of its protagonist Elvis poised to flee Nigeria and enter the United States under a false identity. This uncertain final scene uncomfortably evokes the precarious global mobility of marginal urban migrants whose successive experiences of displacement accumulate to a harmful and alienating degree. As Elvis' identification with the lynched figure in the James Baldwin short story he is reading in the airport departure lounge suggests, his next destination is riven by its own histories of violence and discrimination, which evoke and intensify the hardships that he has already experienced in Lagos.

In *The Beautiful Things That Heaven Bears* (Mengestu 2007a), Dinaw Mengestu portrays the experiences of an Ethiopian exile whose seventeen years in the United States have borne witness to the disappointment of the American dream to which Elvis tentatively clung. Set largely in Washington, D.C. during the early 1990s, the novel is narrated in the first-person by Sepha Stephanos, who fled to America in 1977 at the age of sixteen after his father was murdered by government soldiers during the notorious Red Terror. Orchestrated by Mengistu Haile Mariam, leader of the ruling military government at the time, this violent political campaign brutally repressed those suspected of opposition to his regime.[1] As is revealed midway through the novel, when soldiers find flyers publicising the activist group 'Students for Democracy' during a raid on Sepha's home, his father takes the blame, knowing that his son would be killed if found responsible. Sepha's subsequent flight entails not only the loss of his family, but also the relative privilege of his middle-class upbringing in Addis Ababa. He spends his first year in America living in a high-rise apartment in a poor Maryland suburb with his uncle Berhane, himself a refugee, before deciding to move to Logan Circle in Washington's northwestern quadrant. At that time, the neighbourhood is 'predominantly poor, black, cheap, and sunk in a depression that had struck the city twenty years earlier and never left', but Sepha manages to eke out a living by running a shabby grocery store (Mengestu 2007a, pp. 35–36).[2]

Sepha's only two friends are Joseph and Kenneth, fellow African immigrants from the Congo and Kenya. Together they jokingly refer to themselves as 'the children of the revolution', an initially optimistic moniker that is gradually, tragically ironised by their cumulative experiences of personal and professional disappointment and mistreatment in America.[3] Forced out of their homelands and marginalised within their 'host' country, Sepha and his friends daily experience the 'condition of terminal loss' that Edward Said identifies as that of the 'true exile' (Said 2000, p. 173). By turns nostalgic, homesick and bitter, they regularly meet in Sepha's store where 'inevitably, predictably, [their] conversations find their way home' (Mengestu 2007a, p. 9). The well-trodden paths of these reminiscences fulfil a desire for rootedness that is missing from their daily lives. When remembering the past becomes too painful, they spontaneously interrupt their discussions by testing each other's knowledge of Africa's coups, wars and leaders with reference to an old map of Africa that Sepha keeps taped to his shop wall. The game deflects their longing for home with characteristic dark humour. As Sepha explains, 'no matter how many we name, there are always more, the names, dates, and years multiplying as fast as we can memorise them, so that at times we wonder, half-jokingly, if perhaps we ourselves aren't somewhat responsible' (Mengestu 2007a, p. 8). Elsewhere, Mengestu has written of the damaging impact that uneven Western accounts of Africa's conflicts can have, noting that 'the words "hell" and "horrific" all too often serve as the starting point for a narrative' about the continent (Mengestu 2007, p. 60). For these three refugees, however, irreverently remembering the excesses of such dictators as Bukassa, Amin and Mobutu indicates not the beginning of yet another story about Africa and its supposedly all-encompassing violence, but rather a stand-in for an explanatory narrative. As Caren Irr suggests, 'the terrible, exhausted expertise that this routine creates' is suggestive of 'a stagnant form of immigrant melancholia' (Irr 2013, pp. 50–51). Whenever their conversations demand a full accounting for their respective traumas, the exiles' shared game provides a preferably superficial means of imposing a semblance of order onto the chaos they have left behind and the uncertain situation in which they now find themselves.

While the friends' game speaks to their conflicting emotions about their respective home countries – places that they simultaneously love and despise – they are equally unable to situate themselves within their immediate urban surroundings. Racial and national others, according to the putative norms of American society, they can neither assimilate nor return. Sepha perceives this existential disorientation as a crisis of representation. 'How did I end up here?' he asks himself, 'Where is the grand narrative of my life? The one I could spread out and read for signs and clues as to what to expect next' (Mengestu 2007a, p. 147). His struggle to understand both his geographical setting and historical situation reflects what Fredric Jameson identifies as the 'spatial as well as social confusion' characteristic of contemporary globalisation's disorienting effects (Jameson 1991, p. 54). The 'enlargement of capital', he argues, produces 'a growing contradiction between lived experience and structure' (Jameson 1988, pp. 348–49). As modes of production become increasingly dispersed, so too do the factors that influence individual lives. In turn, the impacts of local, subjective decisions and behaviours are felt further afield than ever before. These disjunctures are perhaps nowhere more intensely experienced than at the margins of contemporary cities, whose disenfranchised populations Zygmunt Bauman describes as 'the human casualties of the planet-wide victory of economic progress' (Bauman 2004, p. 63). Despite this bleak diagnosis, Mengestu refuses to write them off, choosing instead to explore the transformative potential of these embattled urban dwellers throughout his novel.

Faced with the discrepant globalities produced not only by transnational trade and technology, but also conflicts, migration, environmental concerns and cultural connections, Jameson calls for 'an aesthetic of cognitive mapping – a pedagogical political culture which seeks to endow the individual subject with some new heightened sense of its place in the global system' (Jameson 1991, p. 54). In doing so, he builds on insightful earlier work by urbanist Kevin Lynch (1960) who argues that individuals orient themselves within ungraspable urban totalities by identifying and recalling notable buildings, environmental features and other landmarks. These elements allow them to construct a mental image of the city that enables them to move through it physically. Extending the scope of cognitive mapping from individual cities to the world at large, Jameson suggests that the 'alarming disjunction point between the body and its built environment', so keenly felt by multiply displaced subjects such as Sepha, 'can itself stand as the symbol and analogon of that even sharper dilemma which is the incapacity of our minds, at least at present, to map the great global multinational and decentred communication network in which we find ourselves caught as individual subjects' (Jameson 1991, p. 44). In light of this cognitive challenge, Jameson argues, the fundamental role of art is to provide a representational bridge between local and global experience that will enable a better understanding of their mutual imbrication.

Although Jameson deems this new mode of representation to be 'as yet unimaginable', this chapter reveals how Mengestu's novel rises to this challenge by elegantly exploring the conditions that both limit and facilitate our ability to 'grasp our positioning as individual and collective subjects and regain a capacity to act and struggle' in the contemporary global era (Jameson 1991, p. 54). In keeping with Jameson's pedagogical imperative, Mengestu's self-reflexive narrative points to the particularly instructive experiences of those who exist at the global urban margins. Sepha's exilic perspective on the American capital affords an especially useful reconfiguration of what, following Michel de Certeau, can be understood as an ideologically charged 'Concept-city', originally designed to consolidate a unitary national identity (de Certeau 1998, p. 95). Sepha bears critical witness to the gentrification of his historically poor downtown neighbourhood, a paradoxically divisive process that undermines its own professed cosmopolitan intent. If urban redevelopment fails to effectively reimagine the urban space of D.C. in the context of its undeniable globalisation, local efforts to resist unwelcome neighbourhood transformation are similarly hampered by an incomplete understanding of their broader context. By dramatising Sepha's own struggle to cognitively map his individual experience through a symbolic account of local travel, Mengestu offers a more enduring representation of the relationship between the particular struggles of D.C.'s urban margins and the globally widespread mode of order-building from which they are inseparable.

Migrant cognition

By focalising the novel through Sepha's marginal perspective, Mengestu harnesses the particular social and spatial insight afforded by his protagonist's exilic condition. If, as Jameson asserts, 'our insertion as individual subjects into a multidimensional set of radically discontinuous realities' defines contemporary globalisation, an unwilling migrant like Sepha experiences these social and psychic disturbances with particular intensity (Jameson 1988, p. 351). In a well-known essay, Edward Said more optimistically identifies 'originality of vision' as one of the possible 'pleasures of exile' arising from the inevitable 'awareness of simultaneous dimensions' to which displacement gives rise (Said 2000, p. 148). For Sepha, frequent memories of Ethiopia not only interrupt, but largely mediate his experiences in America. While such an expansive worldview can enable a unique perspicacity, Said does also acknowledge that such a positionality is 'both wearying and nerve-wracking' (Said 2000, p. 148). When struggling to concede the loss of his earlier life to the reality of poor migrant existence in D.C., Sepha himself regrets that he 'never could find the guiding principle that relegated the past to its proper place', highlighting the disorienting effect of his reminiscences (Mengestu 2007a, p. 127).

The overall structure of the novel reflects its narrator's fractured subjectivity, conveying to the reader the inherent angst of global cognitive mapping, especially for an exile such as Sepha, who is nevertheless so uniquely placed to do so given the duality of his perspective. The text alternates between various asynchronous and dispersed narratives, indicative of Sepha's temporal and spatial confusion. The immediate action takes place over the course of several days in early May 1993, during which Sepha receives an eviction notice from his rented store that prompts him to take a spontaneous journey around D.C. on foot and by public transport. This storyline is interspersed by Sepha's memories of the personal and political circumstances leading to his flight from Ethiopia, and recollections of his early years in the US. The fourth narrative thread addresses the more recent past of Logan Circle, during which the once blighted neighbourhood has been markedly redeveloped, leading to evictions of many long-standing residents, mostly African Americans, who can no longer afford to pay the increased rents on what were once their homes. Amongst the influx of new affluent white homeowners is Judith, a divorced academic to whom Sepha is tentatively attracted. Charmed by her precocious daughter Naomi, the three form a friendship that comes to an unwelcome end when Judith is forced to relocate after her beautifully restored home is burned down in protest against the gentrification of the neighbourhood.

Mengestu's carefully constructed and explicitly self-reflexive text repeatedly enacts the 'dual thinking' that Jameson posits as essential to formulating effective resistance to the material and social inequities that are so intensified by contemporary global capitalism (Jameson 1988, p. 360). Local activism, Jameson argues, must 'always take place at two levels: as an embattled struggle of a group, but also as a figure for an entire systemic transformation' (Jameson 1988, p. 360). The disjunctive form of the novel reveals Sepha's acute awareness of these different experiential scales. However, his profound sense of loss and displacement suggests the real difficulty of their reconciliation.[4] Moreover, Mengestu demonstrates, the spatial and temporal shifts necessary to the production of a cognitive map exceed the clear duality that provides Jameson with such productive critical purchase. As discussed below, the novel deliberately pluralises 'local space', telescoping back and forth between idealistic city plans, private homes and small commercial spaces. 'Global' space is similarly reconfigured from Sepha's marginal perspective, revealing how currents of violence, desire, memory and grief shape the coordinates of cognitive maps in addition to economic circuits.

Decaying concept, divided capital

The discontinuous structure of Mengestu's novel enacts a form of representational contingency that not only suggests the necessary dynamism of any cognitive map, but also directly undermines the static national ideology inscribed into D.C.'s urban form. Designed in 1791 by the French architect and engineer Pierre L'Enfant and officially established in 1800, Washington, D.C. exemplifies what Michel de Certeau refers to as a 'Concept-city', an ideological space 'founded by utopian and urbanistic discourse' (de Certeau 1998, pp. 94–95). This description has clear resonance with Henri Lefebvre's theory of 'conceived space': ordered, two-dimensional representations of the city, created by 'scientists, planners, urbanists, technocratic subdividers, and social engineers', whose productions of space conceal 'its "real" subject, namely state (political) power' (Lefebvre 1991, pp. 38, 51). As Sarah Luria explains in her fascinating analysis of the relationship between nineteenth-century Washington's architecture, politics and literature, 'discussions surrounding the design of the nation's capital suggest that the planners saw the city as a strategic opportunity for the bodily education of citizens in new political behaviours', including the adoption of a national mindset (Luria 2006, p. xxiv). The grandiose National Mall, for example, both inspires and subdues those who enter and interact with this space through its display of centralised power. Attention-grabbing national landmarks, such as the Capitol, similarly publicise governmental authority to both a domestic and international audience.

Despite the lofty ideals embedded in the capital's early plans, the city itself has always been a contradictory space. In Andrew Holleran's short novel *Grief*, a sensitive portrayal of an unnamed man's temporary residence in the city while mourning the death of his mother, the narrator observes that 'there's something still halfhearted about Washington . . . a city that, block by block, weaves in and out of grandeur and shabbiness' (Holleran 2006, p. 26). This juxtaposition of the 'high' and the 'low' is, de Certeau asserts, inherent to the Concept-city, the realisation of which rests on the attempted 'repress[ion] of all the physical, mental, and political pollutions that would compromise it' (de Certeau 1998, p. 94). As in other idealised cities, such as the morally and materially 'clean' Lagos envisaged by the Nigerian government of Abani's *GraceLand*, the desire for order coincides with the production of waste. That which cannot be successfully eliminated – rundown neighbourhoods, alienated residents, illegal and informal economies – appears and circulates in the Concept-city as notable reminders of its inevitably partial nature. Commenting on the poverty 'just blocks away' from the National Mall, Sarah Luria notes that 'rather than providing an airtight patriotic experience, the capital has an uncanny habit of highlighting the nation's flaws with a kind of confessional zeal' (Luria 2006, p. xxi). Contemporary globalisation further intensifies the representational inadequacies of the Concept-city. National ideology decouples from urban spaces as cities become increasingly important nodes in transnational circuits of migration, business and technology. As Holston and Appadurai note, 'there are a growing number of societies in which cities have a different relationship to global processes than the visions and policies of their nation-states may admit or endorse' (Holston and Appadurai 1996, p. 189). In light of such transformations, the flexible aesthetic of the cognitive map emerges as essential.

Through Sepha, Mengestu charts D.C.'s ideological fault-lines, directing attention to marginal spaces and people that bespeak the 'multiple forms of wretchedness and poverty outside the system and of waste inside it' that de Certeau identifies as the inevitable, atrophic losses generated by the Concept-city (de Certeau 1998, p. 95). Taking the rundown neighbourhood of Logan Circle as its primary setting, the novel directs the city's urban imaginary away from its most well-known buildings, thereby casting a critical light on the exclusive national identity embedded therein. In doing so, he evokes a tradition of Washington writers, such as Edward P. Jones and Marita Golden, who have focused on the daily lives of marginal urban residents rather than the experiences of those central to the city's public political machinations. In his useful anthology of D.C. literature, Christopher Sten describes how

> two traditions – the one local, the other national – have developed side by side, often with a good deal of interplay between them, making Washington writing an unusually rich and resonant, if sometimes schizoid, body of work, with the national government serving as a backdrop, or foil, to the featured lives of local residents.
>
> (Sten 2011, p. 2)

Mengestu's novel, with its deeply transnational perspective, fragments this urban imaginary even more, demonstrating how seemingly 'foreign' places and politics are an equally formative influence on the fabric of Washington life. Just as the Concept-city must necessarily expand to accommodate its undeniable globality, so too must characterisations of the city's literary canon.[5]

Ironically, the closer Sepha moves to the centre of Washington, the more distant he feels from the political ideals enshrined in the city's monumental architecture. When first living with his uncle in Silver Spring, a Maryland suburb north of D.C., Sepha enrolls as a student and enjoys the sense of identity this gives him, which was, he notes with retrospective irony, 'akin to being the citizen of a wealthy foreign country' (Mengestu 2007a, p. 98). However, the casual racism of his part-time employer, the monotony of the back-breaking work he has to perform as a hotel porter and the soul-destroying sight of his dejected uncle 'turn[ing] himself off every morning' in order to endure his own tedious work-days soon dispel 'the liberal idea of America' in which Sepha briefly and naively believes (Mengestu 2007a, p. 98). His decision to move to Logan Circle in downtown D.C. is both an act of self-preservation and silent protest, as he explains:

> When I moved into the neighbourhood I did so because it was all I could afford, and because secretly I loved the circle for what it had become: proof that wealth and power were not immutable, and America was not always so great after all. The neighbourhood, and by extension the city, had fallen, and every night I could see that out of my living-room window.
>
> (Mengestu 2007a, p. 16)

Not only the site of putative moral decline, the figuratively 'fallen' Logan Circle is suggestively described as a territory to be reclaimed in the Concept-city's implicit war against such social contaminants as crime and poverty. During the course of the novel, the view from Sepha's window changes as 'squadlike formations' of workers arrive to begin restoring newly acquired properties, their martial appearance a further indication of the ideological struggle that is playing out in the Circle (Mengestu 2007a, p. 16).

Gentrification, social amnesia and structural control

The self-serving redevelopment of Logan Circle by affluent white newcomers demonstrates a strategic effort to arrest the decay of the national Concept-city. As this parochial urban ideal is increasingly put under pressure by the realities of globalisation, the priorities of city governance must necessarily shift. Beyond the symbolic promotion of a national collective, John Rennie Short argues, a fundamental concern for cities such as D.C. becomes 'the maintaining, securing, and increasing of urban economic competitiveness in a global world' (Short 2004, p. 7). Gentrification plays a key role in this reconceptualisation of urban space, enabling local authorities to maximise their city's ideological and financial capital through the attractive packaging of 'culture, consumption, cool, and cosmopolitan' (Short 2004, p. 7). In Washington as elsewhere, urban branding campaigns seek to attract new downtown residents by emphasising the affordability and social potential of once liminal ethnic neighbourhoods.[6] The goal is that, by renovating dilapidated houses in such areas, new residents raise property prices and secure their own economic clout, transforming neglected neighbourhoods into desirable locales and globally promoting the updated Concept-city as they do so. Thus, de Certeau suggests, city governance attempts to 'transform even deficiencies into ways of making the networks of order denser' (de Certeau 1998, p. 95).

Michael Thompson's *Rubbish Theory* (1979) further illuminates the way in which this kind of gentrification acts as a form of social control. He argues that most objects, houses included, can be classified as either 'transient' or 'durable', the former referring to that which decreases in value over time, such as most mass-produced consumer goods; the latter to that which retains and even increases in value over time, for example unique works of art. Buildings such as those in the Logan Circle neighbourhood, however, belong to a significant third category, that of 'rubbish' (Thompson 1979, p. 7). Such objects have exhausted their value, but not their life-spans. In other words, they 'continue to exist in a timeless and valueless limbo where at some later date . . . [they] have the chance of being discovered' (Thompson 1979, p. 10). Rubbish is therefore characterised by what Thompson calls 'social malleability'; the status of objects that fall within this category has the potential to change according to social demands and pressures (Thompson 1979, p. 10).

The extravagant renovations to which Sepha bears witness dramatise the control mechanism in action as he watches the neighbourhood literally moving from 'decay to respectability' (Mengestu 2007a, p. 190). From the vantage point of his store, he observes 'the plumbers, the electricians, the heating guys, the painters, the roofers, and the architect', who come to repair Judith's house (Mengestu 2007a, p. 16). To a resident like Sepha, who has lived in Logan Circle for a decade without being able to afford the services of any tradesmen, such thorough restoration might seem excessive. However, as Thompson explains:

The amount of maintenance that is deemed reasonable is not a quantity deriving naturally from the intrinsic physical properties of the house and its environment. The level of maintenance that is deemed reasonable for a building is a function of its expected life-span and its expected life-span is a function of the cultural category to which that building is at any moment assigned and, if its category membership changes, so will its expected life-span and its reasonable level of maintenance.

(Thompson 1979, p. 37)

It comes as no surprise that those with access to stronger social networks, financial resources and political power control the assignment of durability and transience to different objects (Thompson 1979, p. 8). In determining Logan Circle worthy of redevelopment, privileged gentrifiers recast its buildings as durable objects, pulling them out of the rubbish category. However, this selective revaluation of urban space does not extend to the neighbourhood's downtrodden residents, ensuring the perpetuation of existing race and class segregation.

In its striving to preserve the idealised space of the Concept-city, gentrification must necessarily flatten the social and historical conditions of its possibility. Thompson suggests that these amnesiac tendencies are characteristic of our relationship to waste in general. As he explains, 'rubbish is always covert, in that we strive quite successfully at all times to deny its existence' (Thompson 1979, p. 20). Its recovery through gentrification is a similarly covert operation in that it too requires a denial, which masks its origins. Like other gentrifiers, Judith acknowledges urban waste when it furthers her personal aesthetic and domestic goals, but she is otherwise content to ignore it, as evidenced by her habit of reading on one of the benches in Logan Circle, 'undisturbed by the drunk men sleeping or stumbling around her' nor the 'whirlwind of fallen leaves and trash that would occasionally rise . . . and flit about in the air as if deliberately calling attention to itself' (Mengestu 2007a, p. 20). Compare her indifference to the neighbourhood's signs of neglect to the behaviour of Mrs. Davis, a widow and long-time resident of the circle who 'in desperate moments of restlessness was known to sweep the sidewalks and street free of litter' (Mengestu 2007a, p. 22). Sympathetic to her loneliness, Sepha explains that 'she was not mad, only bored and looking for the attention of her neighbours' (Mengestu 2007a, p. 22). 'Confident and oblivious to the world', Judith misses the opportunity to engage with her new neighbour, instead inciting a curiosity that quickly becomes hostile (Mengestu 2007a, p. 21). Although, as Jon Binnie notes, 'the global habitus of gentrifiers, superficially at least, seems to reflect the attitudes and practices of cosmopolitanism, including an active celebration of and desire for diversity', Judith's privileged detachment demonstrates how the process 'in fact produce[s] an exclusion of difference by drawing symbolic boundaries between acceptable and non-acceptable difference' (Binnie 2006, p. 16). Although living in a culturally distinct area carries with it a certain cachet for this incoming class of white professionals, their engagement with alterity does not extend beyond their artfully restored homes.

Judith's assured yet selective re-appropriation of rubbish suggests, if not total ignorance, then certainly a damaging naivety regarding what Henri Lefebvre (1991) describes as the social production of space. Not only does this blind-spot prevent her from compassionately relating to her marginalised neighbours, it also ultimately inhibits her relationship with Sepha, a shared attraction that briefly holds out the promise of a mutually fulfilling relationship across Logan Circle's divisive barriers of race and class. Self-aware to some extent, Judith nevertheless enjoys a sense of belonging that Sepha lacks. Drawing on the trope of furniture, so highly prized by the gentrifying classes, Mengestu exposes Sepha's spatial insecurity. Again demonstrating the necessary scalar shifts that cognitive mapping must perform, the narrative emphasises how arranging and furnishing the private realm of the home is central to understanding our position in relation to a public, global space.

When Sepha accepts an invitation to dinner at Judith's house, he is finally admitted to the beautifully restored building that he has admired from afar. His restrictive self-doubt gradually emerges as he observes:

> An old record player and radio the size of a desk, made of wood and with a dozen chrome knobs, sat in the hallway. The living room had a heavy black wall-mounted phone from the early twentieth century, and a silver clock stuck permanently on two-twenty. The leather couches, chestnut colored and densely packed, were separated by a wooden coffee table that had at least fifty small drawers along its side. It was all so solid, comfortable, and familiar, as if Judith had deliberately picked only pieces of furniture that had proven their ability to withstand time.
>
> (Mengestu 2007a, p. 53)

Here, Sepha mistakenly attributes inherent durability to Judith's carefully chosen, but outdated, devices. In fact, her artful recuperation of practically obsolete items such as a broken clock demonstrates her ability to re-categorise technological 'rubbish' as valuable and beautiful. Impressed by Judith's tasteful retro aesthetic, Sepha perceives a timeless beauty in her possessions. Yet these objects do not so much 'withstand time' as evade history. Judith's recovery of these once functional things ignores their past valueless status and the reasons for it.

In contrast to Judith, Sepha's recuperation of material rubbish serves to accentuate its transience and, by extension, his own struggle to secure a permanent sense of belonging in Washington. When he returns from dinner, he is struck anew by the shabbiness and smallness of his own apartment. Although many of his belongings are 'scavenged from the trash', his recycling does not bestow durability upon these secondhand items (Mengestu 2007a, p. 60). Far from fulfilling an idealised retro aesthetic, Sepha's obsolete technology simply doesn't work properly – his old television with 'knob dials and terrible reception' is neither functional nor visually appealing (Mengestu 2007a, p. 60). Similarly,

> The rug in the center of the room had been left by the previous tenant, who had most likely inherited it from the tenant before him. The ends were so frayed that at least twice a month I had to trim a piece off to keep from tripping on the loops of extended thread. Five years later now and one end of the rug was noticeably longer than the other; the corners had been rounded off, and then cut like a pie sliced into at odd, uneven angles.
>
> (Mengestu, 2007a, p. 60)

There is some humour in Sepha's knowing admission of this pointless battle versus the disappearing rug, but the latter's finite life-span underscores the impermanence not only of the apartment's contents, but also their owner's contingent status in the city. His home is literally vanishing from beneath his feet. Whereas Judith, a newcomer, immediately stakes a proprietorial claim to her Logan Circle house owing to her relative wealth, Sepha's residence is, like that of his apartment predecessors, dependent on the demands of his landlord and the property market. Bearing 'the stamp of too many lives and too many people', his furniture signals the many marginalised histories that gentrification erases and ignores (Mengestu, 2007a, p. 60).

Unlike Judith, Sepha is acutely conscious, at all times, of the way in which divisions of race and class are spatially articulated. When Judith does visit Sepha's apartment, she compliments his 'great sense of space', the inevitable outcome, Sepha counters deadpan, of not having anything to fill it with (Mengestu 2007a, p. 85). Sharing a drink together, they raise a toast to 'furniture', after which Judith falls asleep on Sepha's 'hideous couch' (Mengestu 2007a, pp. 86–87). The failure of Sepha's admittedly 'deliberate act of seduction', which he never has the opportunity to retry, can, in a sense, be attributed to the very furniture that they jokingly acknowledge. The contrasting condition of their living room décor makes explicit the social distance that blocks their intimacy.

The ease with which Judith compliments Sepha's bare apartment despite his evident poverty demonstrates her adherence to what Lefebvre calls 'the illusion of transparency' – an illusion that the detailed nature of Mengestu's narrative deliberately contests. Ignoring the fact that '(social) space is a (social) product', Judith maintains that space is 'luminous . . . innocent, free of traps or secret places' (Lefebvre 1991, pp. 27–28). She literally propagates this illusion by continually leaving her house 'fully lit', an extravagant gesture that Sepha perceives as 'distinctly unjust' (Mengestu 2007a, p. 60). Together with the other circle newcomers, their conspicuous displays of wealth proclaim their legitimate, unencumbered presence in the neighbourhood – they have nothing to hide. As Sepha observes:

> It was the same thing with all of the other newly refurbished homes in the neighbourhood; curtains provocatively peeled back to reveal a warmly lit room with forest green couches, modern silver lamps that craned their necks like swans, and sleek glass coffee tables with fresh flowers bursting on top. There was something about affluence that needed exposure, that resisted closed windows and poor lighting and made a willing spectacle of everything. The houses invited, practically begged and demanded, to be watched.
>
> (Mengestu 2007a, p. 52)

All surface and light, these artfully staged interiors demand an audience. Although the narrator of Andrew Holleran's *Grief* asserts that 'there was nothing more agreeable in Washington than walking around circles and houses like these – looking at places other people lived' (Holleran 2006, p. 72), Sepha's own 'window shopping' is somewhat more conflicted (Mengestu 2007a, p. 53). His description of the houses' sensual appeal clearly bespeaks his unrequited desire for Judith, as well as the lifestyle she enjoys. Yet there is something troubling about this un-gratifying form of voyeurism. Sepha explains that 'rarely did I ever see the people who lived in those houses, as if each were merely display-case props of revitalisation' (Mengestu 2007a, p. 53). There is an eerily lifeless quality to this supposed renaissance, which undercuts the initially enticing appearance of the restored houses. Commenting on gentrification's containment of desire, Nikolas Rose explains how the process creates,

> not so much a complex of dangerous and compelling spaces of promises and gratifications, but a series of packaged zones of enjoyment, managed by an alliance of urban planners, entrepreneurs, local politicians and quasi-governmental 'regeneration' agencies.
>
> (quoted in Binnie et. al. 2006, p. 18)

Rather than forging novel alliances, communities and intimacies between their new occupants and poor neighbours, the ostensibly 'transparent' homes around Logan Circle erect new barriers of race and class, thereby eroding the very cosmopolitan promise on which gentrification is founded.

Re-mapping Washington, D.C.

Reduced to spectators while their surroundings change, longstanding residents of Logan Circle anxiously comment that 'the neighbourhood's changing, things are changing, it's not like it used to be, I can't believe how much it's changed, who would have thought it could change so quickly, nothing is permanent, everything changes' (Mengestu 2007a, p. 23). These are, Sepha notes, 'the passive and helpless observations of people stuck living on the sidelines' (Mengestu 2007a, p. 23). Faced with the sudden appearance of ostentatious and seemingly unattainable wealth, it is unsurprising that this marginalised urban population experiences an initial sense of hopelessness. As Michael Thompson notes,

> The operation of [the] control mechanism would seem inevitably to give rise to a self-perpetuating system. . . Those people near the top have the power to make things durable and to make things transient, so they can ensure that their own objects are always durable and that those of others are always transient. They are like a football team whose centre-forward also happens to be the referee; they cannot lose.
>
> (Thompson 1979, p. 9)

Yet while the odds might seem impossibly stacked against the circle's existing residents, Michel de Certeau suggests that resistance to the implementation of the gentrified Concept-city can be found by directing attention to the 'microbe-like, singular, and plural practices which an urbanistic system was supposed to administer or suppress, but which have outlived its decay' (de Certeau 1998, p. 96). The difficulty, as Jameson (1988) notes, is extrapolating from these local struggles a cohesive collective challenge to the uneven system of global capitalism with which gentrification is intertwined. Mengestu dramatises this dilemma through his narration of the varied acts of resistance implemented by both Sepha and his neighbours. While 'these multiform, resistance, tricky and stubborn procedures elude discipline without being outside the field in which it is exercised', the perspectival shifts enabled by Mengestu's fiction allow the reader to newly see that same field as textured and dynamic, indicating the pedagogical promise of his thoughtfully imagined cognitive map (de Certeau 1998, p. 96). Both the content and form of the novel thus reveal the crucially instructive role that the transnational urban margins play in developing an understanding of the global conditions that subtend these local sites of discrimination and struggle.

Sepha can be seen to perform some of what de Certeau terms these 'surreptitious creativities' from within the space of his store (de Certeau 1998, p. 96). Not only does the disillusioned commentary that he offers from behind his register identify the material and social decay of the Concept-city, but his personal and professional decisions actively promote its unravelling. Whereas Kenneth – polite, hardworking and unassuming – plays the role of a model migrant so well that Joseph accuses him, in a moment of drunken cruelty, of behaving like 'the perfect house nigger', Sepha makes no attempt to fulfil such expectations (Mengestu 2007a, p. 182). He is a reluctant shopkeeper, opening his store at odd hours, conceding to shoplifters, ignoring customers, neglecting maintenance and allowing his stock to expire and rot on the shelves. Indeed, he treats his job as an opportunity 'to read quietly, and alone, for as much of the day as possible' (Mengestu 2007a, p. 40). In a useful comparative analysis of conspicuous consumption in Mengestu's novel and V. S. Naipaul's *A Bend in the River*, one of Sepha's favourite books, Dayo Olopade suggests that Sepha's conversion of his shop into 'a hybrid working and living space reclaim[s] the psychic terrain destroyed by the capitalist schism between labouring body and self' (Olopade 2008, p. 149). By refusing the capitalist imperative to accumulate, Olopade argues, Sepha thus reasserts his agency on his own terms. Yet confining analysis to the 'resistance strategies' that Sepha enacts in the space of his store fails to address the broader context for Sepha's ennui (Olopade 2008, p. 149). Sepha's relationship to this local space cannot be understood in isolation from his position relative to America and Ethiopia. His apathy towards his profit margins is symptomatic not only of his disinterest in life as a Washington shopkeeper, but also his disorientation relative to his global circumstances.

By inserting a narrative of local travel into the plot, Mengestu dramatises how migrancy facilitates both the practice and aesthetic of cognitive mapping. The inseparability of Sepha's neighbourhood store from the transnational route that brought him there is made explicit when, like many other long-standing circle residents, he receives an eviction notice due to failure to pay rent on the property. Facing anew the threat of persecution or displacement, Sepha impulsively takes flight, embarking on an unplanned day-trip around the city by foot and on public transport. Unlike his earlier traumatic departure, the strategic repetition of his refugee route enables Sepha to newly understand his physical and social location within a global context. Not only does his itinerary re-map the city itself, exposing the plurality of this dynamic space, but it further reconfigures D.C.'s relationship to Addis Ababa, suggesting previously unexamined connections between nominally distinct 'First' and 'Third' world cities.

Sepha's journey around Washington begins when two middle-aged white tourists enter his grocery store to buy provisions for their self-guided walking tour of the city. Like gentrification, tourism promotes superficial engagement with an idealised city. The practice of sightseeing, as its name suggests, assumes the primacy of optical knowledge: to look at the city is to understand it. They admire the statue of General Logan that can be seen from Sepha's store, but their glorification of this single Civil War hero does not take into account the multiple displacements and acts of violence on which such a selective historical account rests. They are unaware that just one month previously, the statue was 'chipped, defaced, and smeared with human, dog, and bird shit' until the newly formed General Logan Circle Statue Association commissioned its restoration (Mengestu 2007a, p. 36). As Sepha observes, wryly noting the circle's newly imposed standards of moral and material cleanliness, the general now 'looks down on all of us with the glimmering sheen of a privately funded cleaning job' (Mengestu 2007a, p. 36). Like the once omnipresent 'drunk old men', the prostitutes who used to work around the circle have now 'vanished not into thin air, but into a different space or reality' (Mengestu 2007a, pp. 36, 38). These alienated urban residents, treated like the expendable commodities that their customers 'had supposedly left home for', are further displaced by the neighbourhood's supposed 'renaissance' (Mengestu 2007a, p. 38). Their destination unknown, these missing and degraded women haunt the triumphant history announced by General Logan's shining statue.

If the sights that the tourists admire encourage historical amnesia, so too do the 'enormous fold-up maps' that they rely on to navigate the metropolis (Mengestu 2007a, p. 71). The path that the couple follows regulates their physical and, by extension, psychological interaction with their unfamiliar surroundings. De Certeau argues that, by transcribing myriad urban itineraries into legible routes, maps produce 'fixations [that] constitute procedures for forgetting' (de Certeau 1998, p. 97). By following the tourists and embarking on his own impulsive walk around the city, Sepha reverses this process and reinstates the spontaneity of what de Certeau calls 'pedestrian enunciation' (de Certeau 1998, p. 99). Travelling west along P street, for example, the tourists pass 'town homes being built on the left and a two-story organic grocery store being built on the right', sure signs of the neighbourhood's increasing value (Mengestu 2007a, p. 75). Sepha, however, recalls the small-scale businesses that used to occupy this territory. While perhaps not quite as wholesome, the 'grocery store that sold wilted vegetable and grade-D meat, the auto repair shop, and black-owned bookstore called Madame X' served a self-sustaining marginal community (Mengestu 2007a, p. 75). Whereas following a singular cartographic route limits where the tourists go and what they see, Sepha's unplanned journey opens up other pathways through the city and its history, thereby illuminating the 'ensemble of possibilities' concealed by the map's controlled spatial order (de Certeau 1998, p. 98).

The urban knowledge Sepha has acquired over the years both ironises and laments the tourists' naïve perception of the city. As he follows the couple, he reveals his personal disillusionment with the exalted buildings on which they gaze. While the visitors are awestruck by their first glimpse of the White House, Sepha thinks of its presidential occupant as 'a great Santa Claus' (Mengestu 2007a, p. 76). The 'higher power' which is supposed to emanate from the Oval Office has been eroded by his first-hand experiences of Washington's more marginal spaces. Having traveled to the city limits, he knows the 'dirty secret about D.C.': 'For all its stature and statuses, the city could just as easily have been one of the grander suburbs of America' (Mengestu 2007a, p. 101). While the tourists are obediently impressed by the sights, Sepha implies that the White House is not that far removed from what he imagines to be their home: 'a split-level ranch in the suburbs of some midsize city' (Mengestu 2007a, p. 78).

The striking landmarks that the tourists compliantly admire lend what urbanist Kevin Lynch terms 'legibility' to Washington (Lynch 1960, p. 2). His influential study of urban navigation, to which Jameson's conception of cognitive mapping is indebted, posits that individuals find their way through cities by drawing on a 'generalised mental picture of the exterior physical world' that helps them to situate themselves within an otherwise inconceivable urban totality (Lynch 1960,

p. 4). Striking monuments, marked paths and clear boundaries are some of the urban features that facilitate the production of what Lynch eponymously refers to as the 'image of the city.' This subjective mapping procedure resonates with de Certeau's assertion that 'ordinary practitioners of the city', such as Sepha, who 'live "down below", below the threshold at which visibility begins', articulate dynamic and unique versions of urban space that escape conventional cartography by making individual choices about how they interact with their surroundings (de Certeau 1998, p. 93). Lynch further notes the psychic significance of these mental maps, explaining that 'a good environmental image gives its possessor an important sense of emotional security. He can establish an harmonious relationship between himself and the outside world' (Lynch 1960, p. 4). The tourists' visual focus on D.C.'s official architecture reaffirms their presence within a national collective. By sticking to a controlled route through the city, their interpellation into this seemingly coherent totality remains unquestioned.

However, contrary to the unified imaginary that Lynch deems essential to urban harmony, Sepha's mental image of Washington is inherently transnational. In fact, neither Lynch nor de Certeau fully acknowledge the particular challenges that a fragmented, migrant subjectivity poses to urban interaction. When, for example, de Certeau asserts that walkers insert 'a migrational, or metaphorical, city . . . into the clear text of the planned and readable city', he is speaking in figurative terms to suggest how the idiosyncrasies and unpredictabilities of actual urban movement destabilise the Concept-city's cartographic trace (de Certeau 1998, p. 93). Conscious of his own alienated state, Sepha recognises how important it is 'to consider [Washington] not in fragments or pieces, but as a unified whole' (Mengestu 2007a, p. 173). However, he is unable to do so without recourse to memories of his home city of Addis Ababa. Whereas the tourists compliantly observe predictable landmarks, Sepha notes that

> As a capital city, it doesn't seem like much. Sixty-eight square miles, shaped roughly like a diamond, divided into four quadrants, erected out of what was once mainly swampland. Its resemblance to Addis, if not always in substance, then at least in form, has always been striking to me. As a city, Addis wasn't much larger. Ninety square miles, most of which was a vast urban slum built around the fringes of a few important city centers. The two cities share a penchant for circular parks and long diagonal roads that meander and wind up in confusion along the edges. Even the late-afternoon sun seems to hit D.C. the same way.
>
> (Mengestu 2007a, pp. 173–74)

In keeping with Jameson's assertion that subjective urban navigation is analogous to global cognitive mapping, the itinerant formation of Sepha's expansive mental image of D.C. suggests the inseparability of these two processes.

Sepha's unplanned journey around D.C. suggests important linkages between the American capital and its Ethiopian counterpart, which have been neglected by dominant world city discourse.[7] Whereas existing frameworks of urban comparison which have tended to divide the study of cities into hierarchically opposed 'First-' and 'Third-world' spaces, Sepha's memories imply a discomforting parity between Addis Ababa and Washington, D.C. Not only does this further trouble the coherence of the national Concept-city, it also deepens Mengestu's critique of gentrification's exclusionary social violence.

Sepha's imaginative excursions to his home city of Addis Ababa imply that official power structures are equally corrupt in both cities, undermining the moral exceptionalism implied by D.C.'s official architecture. After Sepha gives the tourists 'an enthusiastic wave good-bye', he lies on the grass in Dupont Circle surrounded by office workers having lunch (Mengestu 2007a, pp. 78, 91). He hears the sound of approaching sirens and, within seconds, a 'parade of police motorcycles, cars, massive black SUVs, and black limos' appears (Mengestu 2007a, p. 92). As audibly intrusive as the Capitol is visually impressive, this disturbing spectacle of power instantly reminds Sepha of Ethiopia where 'troops used to line whatever route the emperor took hours in advance. They swept the street clean of beggars, cripples and trash, and had faithful loyalists stand on the side of the road, ready to bow as he passed' (Mengestu 2007a, p. 93). Despite this insight, Sepha feels an immediate affinity with the other mesmerised observers who 'all have the sense that someone of great import is passing' (Mengestu 2007a, p. 93). When the police cars disappear, Sepha goes on to imagine 'an entirely empty motorcade whose sole purpose is to remind people of what they are up against' (Mengestu 2007a, p. 93). If, as Achille Mbembé has argued, a notable feature of the postcolonial state is its tendency towards excessive spectacle even in the most mundane of circumstances – what Mbembé terms the 'banality of power' – Sepha's observation makes it clear that such posturing is not restricted to sub-Saharan Africa alone (Mbembé 2001, p. 102). Sympathetic to the plight of those marginalised Addis residents who are literally treated like waste to be removed from sight and mind, Sepha's memory implies a current of transnational urban solidarity that persists despite the repressive power structures common to both locales.

Sepha's encounter with the motorcade reveals a structural connection between Addis Ababa and Washington, D.C. that extends beyond mere formal resemblance. But more disturbing than this common symbolic expression of power is the continuum of actual state violence that links both cities. Towards the end of his journey, Sepha reflects on the physical resemblance between Logan Circle and a small park in Addis where he used to take walks with his father. Originally a place for them to escape from the bustle of urban life and 'block out the world in order to live quietly for half an hour or so with [their] thoughts', this sanctuary is shattered during the Red Terror when 'seven bodies [are] neatly lined up in the center of the grass' as a warning to other potential 'traitors' to Mariam's revolution (Mengestu 2007a, p. 217). The figurative human trash swept from the streets in order to make way for a presidential motorcade is here rendered literally abject. The haunting of Logan Circle by this site of violence is, of course, symptomatic of Sepha's personal trauma. As a teen, he was no doubt profoundly disturbed by this shocking encounter with the Ethiopian state's easy elimination of boys his own age who had supposedly dared to question the government's authority. Yet this irrepressible memory further suggests the implicit violence of the gentrification that his present neighbourhood is experiencing. Sepha's transnational cognitive map, almost complete by this stage in the novel, thus recontextualises the redevelopment of Logan Circle as an act of state violence, albeit different in kind to that Sepha experienced in Addis. Viewed from this perspective, Mengestu again emphasises that gentrification needs to be seen not only as a strategic manoeuvre with decided local effects, but also a symptom of a globally widespread mode of order-building that rests on the violent elimination of wasted urban populations.

Rumours, resistance, reconciliation

For Sepha, the connection between the struggles of his new neighbourhood and that which he unwillingly left behind in Addis Ababa are made explicit when Mrs. Davis brings him a 'stack of flyers' to distribute in protest against Logan Circle's gentrification (Mengestu 2007a, p. 194). Reminiscent of those 'inconsequential' student flyers that had provided the catalyst for his father's beating and presumed death at the hands of the Ethiopian military, Sepha notes: 'I knew that there were patterns to life, but what I had never understood until then was how insignificant a role we played in creating them' (Mengestu 2007a, p. 126, p. 194). Here, he acknowledges the totality of the global system that yokes together urban violence in both Addis and D.C., but, as his somewhat defeatist response to this insight suggests, organising effective communal resistance to such oppression presents seemingly insurmountable challenges.

Displaying an energy that Sepha lacks, Mrs. Davis and 'the other widows of the neighbourhood' instigate a meeting of the newly formed Logan Circle Community Association with the intention of developing a plan of action against the illegal evictions that have been taking place, which includes gathering signatures for a petition to the city council (Mengestu 2007a, p. 195). Portrayed from Sepha's jaded viewpoint, the sparse audience that gather to hear Mrs. Davis's 'rehearsed and scripted' speech in a cold church basement seem unlikely to achieve the results they desire (Mengestu 2007a, p. 198). Yet, through his portrayal of the 'grievances and frustrations' expressed during the meeting, Mengestu suggests that the ultimate failure of the community association is not a question of political experience or aptitude, but rather an inability on their part to imagine the true contours of their struggle; in other words, a failure to cognitively map their situation (Mengestu 2007a, p. 199). Unlike Sepha, who enters the meeting with the insight painfully gained through coerced migration, the longstanding residents of Logan Circle cannot precisely articulate the actual causes of their unwelcome change in circumstances. Understandably preoccupied with their immediate security and safety, they nevertheless construct this particular dilemma in more general terms, as Sepha notes: 'The grievances and frustrations came quickly. Some had to do specifically with changes in the neighbourhood, others were more general and came from a deeper, longer-standing frustration with life' (Mengestu 2007a, p. 199). In the absence of a cognitive map that would allow them to better understand the relationship between these seemingly disparate concerns, the residents turn to generalising those responsible for their newly precarious living conditions. Sepha recalls,

> I don't know who used the word 'they' first. . . . Once the word entered the meeting, it seemed to trail onto the end of every sentence. I don't know who they think they are. What are they doing here anyway. They have their own neighbourhoods and now they want ours too. It's bad enough that they have all the jobs and schools. I was convinced that if given enough space and time, a conclusion would have been drawn that held 'them' responsible not only for the evictions in the neighbourhood, but for every slight and injury each person in that room had suffered, from the children who never made it past junior high to the unpaid heating bill waiting in a dresser drawer.
>
> (Mengestu 2007a, p. 200)

The indignant tension in the room is exacerbated by Judith's presence at the meeting. The only white person there, she quickly becomes the target of her neighbours' anger, being told to 'shut up' when she ventures to express her opinion (Mengestu 2007a, p. 200).

If, as previously mentioned, one of the critiques of gentrification is that, despite its professed desire for diversity, it actually promotes discrimination, the community association meeting reproduces these imposed prejudices. Judith and her fellow newcomers undoubtedly show a superficial appreciation for the social and cultural distinctiveness of Logan Circle, but their engagement with the conditions of its production extends only to property restoration, not community rebuilding. They are only interested in buildings, not people. However, the community organisers are also selective about which social boundaries should be breached. Here, Sepha's critical outsider's perspective is once again crucial. He, like Judith, is an outsider, who 'had snuck into the neighbourhood' to take advantage of its affordable rent at the time (Mengestu 2007a, p. 189). His presence at the meeting is tolerated on the basis of a superficial racial identification that, in this instance, inhibits rather than advances effective local protest.

The community association's rapid circulation of accusations against an amorphous racialised enemy evokes Jameson's assertion that 'conspiracy plots' are 'the poor person's cognitive mapping in the postmodern age' insofar as they substitute 'sheer theme and content' for a fully realised account of the totality in which such inequities occur (Jameson 1988, p. 356). The Logan Circle residents are clearly not just being paranoid – their livelihoods are at stake and they have good reason to understand this persecution in the historic context of African American oppression in D.C. specifically and in America as a whole. Race and class have long intersected to marginalise these urban residents. Yet by failing to give a specific form to their oppressors, characterising the parties responsible simply as 'they', the community association members hamper the creation of a truly effective plan of opposition. They can only conceive of solutions to their problems in similarly general terms. When someone begins to throw bricks through the windows of property owned by newcomers (Judith's car; the lobby of an apartment building), hearsay substitutes for a detailed narrative of resistance. The residents begin to hear and spread 'rumors of marauding men in black touring through the neighbourhood', formulating yet more conspiracies (Mengestu 2007a, p. 219). When the perpetrator of the brick-throwing is revealed to be 'only one desperate, lonely man', the narrative of resistance quickly expires, as Sepha explains: 'Following Frank's arrest, the marauding men in black retreated to the corners of the imagination that had created them' (Mengestu 2007a, p.224, p. 226). Lacking the detail of Mengestu's own nuanced fiction, these stereotypes are a temporary imaginative fix for an enduring representative problem. Instead of enabling productive social action, they offer short-lived consolation for the Logan Circle residents.

Although Joseph anticipates the emergence of 'an entirely new neighbourhood' in the wake of these micro-protests against gentrification, towards the close of the novel Sepha observes that 'in the end, nothing changed' (Mengestu 2007a, p.224). Judith and Naomi leave the circle and with them the promise of romance and friendship, newcomers continue to move into the neighbourhood and older residents such as Mrs. Davis still live with the threat of eviction. Jameson offers the consolation that,

> successful spatial representation today need not be some uplifting social–realist drama of revolutionary triumph but may be equally be inscribed in a narrative of defeat, which sometimes, even more effectively, causes the whole architectonic of postmodern global space to rise up in ghostly profile behind itself, as some ultimate dialectical barrier or invisible limit.
>
> (Jameson 1988, pp. 352–53)

The novel's concluding scene, which sees Sepha admiring his store from the steps of Judith's old house, explicitly stages this global haunting of local space. Directly addressing his dead father, Sepha offers an aphorism of his own to match those his parent used to pass onto him: 'Father: a man stuck between two worlds lives and dies alone. I have dangled and been suspended long enough' (Mengestu 2007a, p. 228). Unlike Chris Abani's Elvis, whose social and spatial orientation remains uncertain at the end of *GraceLand*, Sepha's plot concludes with a degree of resolution. Having acknowledged the global extent of his displacement – the 'two worlds' he experiences – he observes with great precision that 'there are approximately 883 steps between these steps and my store. A distance that I can sprint in less than ten seconds, walk in under a minute' (Mengestu 2007a, p.228). With the stylistic deftness demonstrated throughout the novel, Mengestu here switches spatial scales to reveal how Sepha can now, having completed his instructive journey around D.C., start to reconcile his immediate experience to its broader context. Infused with loss as the acknowledgement of his father suggests, this concluding scene can nevertheless be read as a productive moment of utopian projection. 'Right now', Sepha asserts, 'I'm convinced that my store looks more perfect than ever before' (Mengestu 2007a, p. 228). While by no means a story of unmitigated 'revolutionary triumph', Mengestu's narrative succeeds where those of both the neighbourhood's gentrifiers and its longstanding residents fail. Whereas they both project limited visions of Logan Circle that variously deny the true causes and content of this urban space, Sepha's marginal perspective lends nuance to the neighbourhood itself, the city of Washington, D.C. and its transnational relations. *The Beautiful Things That Heaven Bears* thus successfully dramatises and enacts global cognitive mapping, revealing through the scalar shifts of its content and form the inherent pain and instructive potential of this imaginative endeavour.

Notes

1 In *A History of Ethiopia* (1994), Harold G. Marcus notes the 'unspeakable horrors' perpetrated during this period, which forced 'thousands of Ethiopia's best-educated and idealistic young people' into exile (p. 196). Nega Mezlekia's *Notes from the Hyena's Belly* (2000) offers a compelling non-fiction account of the eruption of this 'political volcano', which sees the author's initial support for the revolution falter in the face of what he describes as a 'calamity beyond anyone's wildest imagination' (pp. 293–94). Maaza Mengiste also explores this violent political transition in her novel *Beneath the Lion's Gaze* (2010).

2 While there are certainly autobiographical influences on the novel – Mengestu's own parents left Ethiopia for America in 1978 when he was two years old – he has repeatedly resisted the conflation of his protagonist's experience with his own, explaining in an interview with *The Rumpus* (Mengestu 2010a) that he is 'always troubled when people want to know what's real in fiction, to parcel things out into the parts that are true and not true'. He wrote *The Beautiful Things That Heaven Bears*, his debut novel, before returning to Ethiopia for the first time at the age of twenty-five, an experience he describes in 'Returning to Addis Ababa' (Mengestu 2010b).

3 The first UK edition of the novel was published under the title *Children of the Revolution* in 2007.

4 Mengestu himself is similarly wary of critical attempts to reconcile the inherent tensions of his work by labelling it as 'immigrant literature', a category that he finds both marginalising and immobilising. His work, he asserts in an interview with *The Paris Review*, is 'American and African at all points and times', demanding its readers' continual engagement with disparate geographies, experiences and traditions (Mengestu 2010c).

5 In his foreword to *Literary Washington*, Alan Cheuse observes that 'part of this peculiar city's sense of place is that it serves as a capital for people who have no permanent sense of place' (Cheuse 2012, p. vii). It is notable, therefore, that neither this nor Sten's anthologies excerpt contemporary migrant writing about Washington. Mengestu is absent, as is V. S. Naipaul, whose short story 'One Out of Many' (Naipaul 1971) offers an unsettling account of an Indian domestic's migration from Bombay to D.C. with his wealthy employer, which places him as a critical witness to the city's 1968 riots.

6 See Gibson (2006) for a useful analysis of the 'City Living, DC Style' initiative launched by Washington Mayor Anthony Williams in 2003, in which the author argues that such 'campaigns are best viewed as a form of semiotic warfare pitched against an amorphous enemy: the image of 'urban decay' that has monopolised American discourse about the city since at least the 19th century' (Gibson 2006, p. 261).

7 Jennifer Robinson's *Ordinary Cities* (2006) offers a notable challenge to these divisive existing paradigms, in which she 'proposes a post-colonial urban studies in which scholars of wealthy, Western cities learn about their cities by thinking with scholars and artists from other places' (Robinson 2006, p. xi).

References

Bauman, Z. (2004) *Wasted Lives: Modernity and its Outcasts*. Cambridge: Polity Press.

Binnie, J. et. al. (eds.) (2006) *Cosmopolitan Urbanism*. London: Routledge.

de Certeau, M. (1998) *The Practice of Everyday Life* (trans. S. Rendall), Berkeley, CA: U of California P.

Cheuse, A. (2012) 'Foreword', in Allen, P. (ed.) *Literary Washington, D.C.* San Antonio, TX: Trinity UP.

Gibson, T. A. (2005) 'Selling City Living: Urban Branding Campaigns, Class Power, and the Civic Good', *International Journal of Cultural Studies*, vol. 8, no. 3: pp. 259–80.

Holleran, A. (2006) *Grief*. New York: Hyperion.

Holston, J, and Appadurai, A. (1996) 'Cities and Citizenship', *Public Culture* vol. 8, no. 2: pp. 187–204.

Irr, C. (2013) *Toward the Geopolitical Novel: U.S. Fiction in the Twenty-First Century*. New York: Columbia UP.

Jameson, F. (1988) 'Cognitive Mapping', in Nelson, C. and Grossberg, L. (eds.) *Marxism and the Interpretation of Culture*. Chicago, IL: U of Illinois P: pp. 347–60.

Jameson, F. (1991) *Postmodernism: Or, the Cultural Logic of Late Capitalism*. Durham, NC: Duke UP.

Lefebvre, H. (1991 [1974]) *The Production of Space* (trans. from French by D. Nicholson-Smith), Oxford: Blackwell.

Luria, S. (2006) *Capital Speculations: Writing and Building Washington, D.C.* Durham, NH: UP of New England.

Lynch, K. (1960) *The Image of the City*. Cambridge, MA: MIT Press.

Marcus, H. G. (1994) *A History of Ethiopia*. Berkeley, CA: U of California P.

Mbembé, A. (2001) *On the Postcolony*. Berkeley, CA: U of California P.

Mengestu, D. (2007a) *The Beautiful Things That Heaven Bears*. New York: Riverhead

Mengestu, D. (2007b) 'Children of War', *New Statesman* 18 June: pp. 60–61.

Mengestu, D. (2010a) Interview by Anne Shulock. *The Rumpus,* 19 October [Online]. Available at http://therumpus.net/2010/10/the-rumpus-interview-with-dinaw-mengestu/ (Accessed 17 February 2016).

Mengestu, D. (2010b) 'Returning to Addis Ababa', *Callaloo*, vol. 33, no. 1: pp. 15–17.

Mengestu, D. (2010c) Interview by Thessaly La Force. *The Paris Review*, 28 October [Online]. Available at http://www.theparisreview.org/blog/2010/10/28/dinaw-mengestu/ (Accessed 17 February 2016).

Mengiste, M. (2010) *Beneath the Lion's Gaze*. New York: W. W. Norton.

Mezlekia, N. (2000) *Notes from the Hyena's Belly: An Ethiopian Boyhood*. New York: Picador .

Naipaul, V. S. (1971) 'One out of Many', in *In a Free State*. London: André Deutsch.

Olopade, D. (2008) 'Go West, Young Men: Conspicuous Consumption in Dinaw Mengestu's *The Beautiful Things That Heaven Bears*, as Prefigured by V. S. Naipaul's *A Bend in the River*', *Transition* vol. 100: pp. 134–51.

Robinson, J. (2006) *Ordinary Cities: Between Modernity and Development*. London: Routledge.

Said, E. (2000) *Reflections on Exile and Other Essays*. Cambridge, MA: Harvard UP.

Short, J. R. (2004) *Global Metropolitan: Globalizing Cities in a Capitalist World*. London: Routledge.

Sten, C (ed.) (2011) *Literary Capital: A Washington Reader*. Athens: U of Georgia P.

Thompson, M. (1979) *Rubbish Theory: The Creation and Destruction of Value*. Oxford: Oxford UP.

4   Seeing the obvious?

Contradictory visibilities in Indian City literature

Bombay's contradictory visibilities

In a memorable early scene from Danny Boyle's 2008 hit movie *Slumdog Millionaire*, set in contemporary Bombay, the young protagonist Jamal is trapped in a filthy public outhouse when the helicopter of the movie star Amitabh Bachchan lands nearby.[1] So desperate is the five-year-old slum-dweller to procure the autograph of his idol that he escapes by plunging into the stinking cesspit below the latrine. Covered from head to foot in human excrement, he trudges towards his hero, successfully parting the assembled crowd of clamouring fans in order to gleefully obtain the actor's coveted signature.

Bachchan himself does not actually appear in the film, but he was one of many observers to comment on the rapturous global reception of its vivid portrayal of Indian poverty.[2] A widespread critical and commercial success in Europe and America, the film's Western embrace was signaled by the eight Oscars it won at the 2009 Academy Awards. However, some viewers raised concerns not only about the production's potentially exploitative use of inexperienced child actors, but also its dismissive treatment of the actual hardships faced by Bombay's slum-dwellers.[3] Although the film itself mocks the naivety of European and American tourists regarding India, it arguably turns deprivation to diversion in comic scenes such as the above. Whereas, for Boyle, an encounter between a shit-covered slum kid and a god-like movie star provides humour through contrast, the absence of adequate toilet and sewage facilities in slums such as Dharavi, where much of the filming took place, is a massive ongoing affront to the health and dignity of the city's poorest residents. Despite Boyle's own longstanding fecal fixation – a notably Pynchonesque scene from his earlier *Trainspotting* (1996) features the heroin-addicted protagonist Renton diving into 'Scotland's worst toilet' to retrieve his stash – he seems bemused by the practical and personal problems that Bombay's lack of toilets presents. Discussing his time spent filming in India, he recalls that

> There's nowhere to shit; people shit everywhere. Although you never see the women shitting. . . You see men doing it all the time. Men and boys. All the time – and you have to get your head around that. But you never see women.

There were all these rumours: 'Oh, they get up in the night' – but I was up in the night, and I never saw them.

(Film4 n. d.)

Boyle's failure to see the female slum-dwellers who must, of course, find time and space to relieve themselves, suggests the particular difficulty that women face in maintaining the personal privacy that is routinely denied to the urban poor. As Mike Davis notes, 'being forced to exercise body functions in public is certainly a humiliation for anyone, but, above all, it is a feminist issue' (Davis 2005, p. 140). A lack of toilets does not just expose slum women to disease and embarrassment, but also harassment and assault.

The ironic distance between Boyle's toilet humour and the uncomfortable reality from which it takes inspiration demonstrates the contradictory dynamics of visibility to which Bombay's slum-dwellers, especially women, are subject. *Slumdog Millionaire* joins a recent spate of films, TV shows, exhibitions and books in exposing Dharavi to a global audience.[4] Internationally recognised as the signs of an 'exotic' poverty, the residents of Bombay's slums are nevertheless, as Arjun Appadurai notes, 'socially, legally, and spatially marginal' within the city and the nation-state (Appadurai 2002, p. 35). Both global signifiers and 'invisible citizens', they experience multiple discrepant degrees of exposure, which render them vulnerable in different ways (Appadurai 2002, p. 35).

In contrast to the reductive images of the slum purveyed by recent artistic and academic works, local popular protests have sought to reassert control over how and by whom Bombay's urban poor are seen. Appadurai praises the 'toilet festivals' organised by the Mumbai Alliance, a coalition of the city's grassroots urban activists who invite state and World Bank officials to celebrate the inauguration of public toilets as part of their broader commitment to achieving effective political recognition of India's slum populations (Appadurai 2002, p. 39). The inability to separate oneself from one's own waste conditions poverty, resulting in the degrading conflation of marginalised people with their own shit. As public spectacles, the toilet festivals 'are a transgressive display of this fecal politics', in which 'humiliation and victimisation are transformed into exercises in technical initiative and self-dignification' (Appadurai 2002, p. 39). As Gay Hawkins notes in her related analysis, 'by refusing to shit in public, members of the Mumbai Alliance use waste to create a private personhood and with it a nascent citizenship' (2006, p. 68). Whereas a young slum-dweller's willing immersion in shit is offered up for comic effect in *Slumdog Millionaire*, the toilet festivals effect a celebratory reversal of this naively self-degrading gesture. By taking control of how and where they produce, dispose of and display their bodily waste, these slum activists importantly differentiate between their private lives and their public exposure – a highly personal problem thus forms the basis for significant collective action. By doing so, they claim an individual agency that merits social recognition and state protection. Here, urban waste practices define and test Bombay's contradictory dynamics of visibility, demanding more equitable and respectful treatment of the urban margins.

Mehta's marginal vantage point

This chapter examines a text that attempts to harness the performative energy of the toilet festivals for a sustained literary critique of the contradictory dynamics of visibility that intersect the margins of contemporary Bombay. Whereas the actions of the Mumbai Alliance foster generative instances of spectacular slum solidarity, Suketu Mehta's *Maximum City* (2004) contextualises the need for such interventions while addressing a broader, transnational audience. Widely praised on publication for its lively content, absorbing detail and elegant style, this work of creative non-fiction provides a kaleidoscopic account of those people and places that exist at the jarring interface of material exposure and social invisibility in Bombay. Mehta's compelling three-part analysis of the 'power', 'pleasure' and 'passages' that shape the margins of the city combines reportage with memoir, ethnography and travel writing to produce a fragmented, multi-genre text that embodies metropolitan kinesis. An experienced journalist, Mehta draws on interviews, observation and lived experience to inform his various portraits of slum-dwellers, gangsters, political activists, bar dancers, movie stars and urban migrants. In keeping with the fecal politics asserted by the Mumbai Alliance, he reveals Bombay's urban waste, in the form of its polluted places and their dejected occupants, to be an important testing ground for new forms of citizenship in contemporary Bombay.

Whereas Dalit literature in Marathi, Hindi and other Indian languages has a long tradition of representing marginal urban existence across genres, *Maximum City* marks a new departure for Anglophone Bombay writing, which has in the past focused on the introverted dramas of the urban middle class.[5] In Shashi Deshpande's *That Long Silence* (1988), for example, the first-person narrator Jaya feels an affinity with her poor female relatives and domestic maids, but her own middle-class crisis of identity ultimately supersedes the struggles of these working women. More recently, Aravind Adiga's *Last Man in Tower* (2011) offers a comic fictionalisation of the threat that urban redevelopment poses to the proudly middle-class residents of a northern Bombay apartment block. Mehta is likewise concerned with the shifting physical and social characteristics of the city in the twenty-first century, but he assumes the distinctly marginal vantage point of Bombay's 'wasted' people and places in order to examine them.

In *Maximum City*, Mehta's self-reflexive narrative of Bombay goes beyond urban explanation to encourage critical reflection on the ways in which the city comes to be known. He does not simply illuminate its frequently misrepresented margins, but he also interrogates different perceptions of them. As discussed below, his sustained engagement with the Bollywood imaginary lends the text a distinctive cinematic optic, but he avoids lazy appropriation in favour of exploring the kaleidoscopic interactions of various illusory Bombays with their material counterparts. In addition to popular cinema, he further draws on a literary tradition of migrant writing about Indian urban waste, which, in light of the city's recent seismic shifts, invites an updated account of Bombay's margins and suggests strategies for doing so. Assuming a multi-focal narrative perspective, Mehta reveals the deep structural and historical interconnectedness of the city, which is denied by its paradoxical visibilities. A deliberately self-conscious writer, Mehta models his own inseparability from the marginal subjects of his text through his adoption of an engaged ethnographic narrative method. However, although he successfully describes and critiques the contradictory dynamics of visibility that shape life at Bombay's margins, his inescapably gendered perspective also unwittingly perpetuates them.

The Bollywood imaginary

Given Bombay's significance as both the dramatic setting and site of production for numerous popular Hindi films, it is perhaps unsurprising that criticism of *Maximum City* has to date focused on Mehta's formal and thematic allusions to Indian cinema. Like recent novels that embrace the suspense of Bombay's nightlife and underworld, such as Gregory David Roberts's *Shantaram* (2004) and Vikram Chandra's *Sacred Games* (2007), *Maximum City* has earned criticism from some quarters for what has been perceived as its melodramatic, Bollywood-inflected portrayal of Bombay's crime, corruption and systemic violence. Mukund Belliappa, for example, argues that Mehta reproduces 'sloppy myths about the city' in his reliance on testimony from 'stock characters, [who are] listened to uncritically and portrayed without nuance' in his account of the city's margins (Belliappa 2008, pp. 348, 358). Favourable readings of Mehta's narrative, of which there are many – *Maximum City* earned Mehta a Pulitzer Prize nomination in 2005 – also identify the merits of the book with its 'tactile sensationalism', suggesting that Mehta's engagement with cinematic paradigms is an inevitable response to confronting the reality of Bombay (Mazumdar 2007, p. 149).[6]

The ready dismissal of Mehta's references to Bollywood narratives does a disservice to the complex cultural critique offered by both popular film and *Maximum City*. A striking feature of recent Bollywood scholarship is the suggestion that these visual texts not only provide an entertaining escape from the demanding realities of South Asia's pre-eminent mega-city, but that they also facilitate urban existence by providing explanatory frameworks for those who daily confront Bombay's challenging mixture of cultural creativity, infrastructural weaknesses, threatening violence and social transformation. In an important essay on the analogous relationship between Indian cinema and Indian slums, Ashis Nandy identifies the crucial political education offered by the former, noting that 'popular cinema not merely shapes and is shaped by politics, it constitutes the language for a new form of politics' (Nandy 1998, p. 12). In a more recent essay, Vyjanathi Rao, makes the similarly compelling suggestion that 'the cinematic might be more than a mere lens – it may also serve as a specific way of engaging with the city', asserting that it is best understood as a 'perceptual technology' that can render unpredictable city existence legible to a broad audience (Rao 2011, p. 9).

Alert to the formal possibilities of Hindi film, Mehta finds in cinematic fragmentation an appropriate mode through which to capture the excess and intensity of contemporary Bombay, a city that he initially finds difficult to grasp. *Maximum*

City starts out as the personal memoir of a returning migrant eager to rediscover the 'city that has a tight claim on [his] heart' (Mehta 2004, p. 3). Having emigrated to America at the age of fourteen, Mehta has cultivated, in exile, a longing for the country of his birth and the city of his childhood, which is shattered by the reality of return. The memories of Bombay that he has nurtured in absentia turn out to be, like those of another well-known Bombay emigrant, 'fictions, not actual cities or villages, but invisible ones, imaginary homelands, Indias of the mind' (Rushdie 1991, p. 10). Just as Salman Rushdie turns his expatriate's alienation to authorial advantage in *Midnight's Children* (1981), making of the past's irretrievability a compelling theme and spur to formal innovation, Mehta likewise responds to the elusiveness of past and present Bombays with stylistic experimentation. Although he declares his personal quest at the very beginning of the text, it does not follow a well-defined linear trajectory, in much the same way that Hindi cinema, according to film scholar Rosie Thomas, places 'emphasis on emotion and spectacle rather than tight narrative . . . on a succession of modes rather than linear denouement' (Thomas 2008, p. 29). Autobiography joins journalistic reportage, oral anecdotes, historical narration and occasional storytelling in *Maximum City* to produce an impressively composite text that shares what Thomas describes as the 'slippage between registers' characteristic of Bollywood films, most notable in their use of what can seem – to some uninitiated Western audiences at least – discontinuous song and dance routines (Thomas 2008, p. 28). Mehta identifies in Hindi cinematic form a rhetorical invitation to polyphony and generic multiplicity that allows him to better explore the multi-dimensionality of Bombay's urban margins.

Although popular cinema provides Mehta with an important navigational tool, urban scholars Thomas Blom Hansen and Oskar Verkaaik caution that 'some urban spaces are so heavily mythologised and enframed through circulating images and narratives that they suffuse, if not overdetermine, any empirical or sensory experience' (Hansen and Verkaaik 2009, p. 6). Indeed, many Indians throughout the nation and diaspora come to know Bombay best through Bollywood movies. Although popular gangster films such as *Parinda* (1989), *Satya* (1998) and *Company* (2002) introduce Bombay's strugglers, slum-dwellers and criminals to a broad audience, there is a risk that such films supplant urban actuality with urban fantasy, in contrast to Indian art-house films such as Mira Nair's *Salaam Bombay!* (1998), which document the hardships of marginal urban existence in a more realist fashion. In a manner not dissimilar to *Slumdog Millionaire*, popular cinematic representations of Bombay thus intensify the paradoxical visibility of the city's margins. The contrast between the sizeable cinema-going audiences they attract and their subjects' actual lack of acknowledgement by the state highlights the disjunctive value accrued by urban waste within discrepant cultural and social economies.

Rather than simply appropriating popular cinematic themes and tropes as some have claimed, Mehta therefore offers a critical engagement with the Bollywood imaginary throughout *Maximum City*, deliberately highlighting both its revelatory promise and concealing effects. He frequently notes discrepancies between the world of the cinema and the 'real world', telling the reader, for example, that the professional criminals he meets 'don't look anything like their movie counterparts' (Mehta 2004, p. 165). In his dramatic eyewitness account of a violent police interrogation, he further signals the gap between imaginary and actual experiences of life at Bombay's margins. While conducting research for a movie script he is working on, he visits a police station with his friend the film director Vinod Chopra, and Chopra's wife Anu. Two counterfeiters who have been caught driving a car carrying hundreds of fake 500-rupee notes are being questioned. With pace and immediacy, Mehta describes in detail how two policemen beat the accused all over their bodies and faces with their own belts and 'a thick leather strap, about six inches wide, attached to a wooden handle' (Mehta 2004, p. 151). The recently appointed Additional Commissioner of Police, Ajay Lal, whom Mehta describes as having the 'towering good looks' and charisma of a movie star, oversees proceedings, liberally cursing the criminals and barking orders at his team, which Mehta quotes directly for full impact (Mehta 2004, p. 132). Eventually, the two counterfeiters are dragged from the office in order to be electrocuted out of sight.

When leaving the police station, a shocked Anu remarks that 'There's a whole world around us that we know nothing about. . . . I just want to watch my Hindi films and be safe' (Mehta 2004, p. 153). By including this admission, Mehta allows an ironic joke at the expense of himself and his similarly bourgeois friends. His acknowledgment of his own privilege versus the deprivations of the marginal urban figures he encounters is characteristic of his self-aware narrative style. However, beyond deflecting potential criticism of their privileged standpoint, Anu's acknowledgement of the gap between the cinematic entertainment that she enjoys (and her own husband purveys) and the visceral scene they have just witnessed suggests the distance between the world of the movies and the reality of Bombay. It is this disjuncture that Mehta seeks to bring to the fore in his engagement with Bollywood in order to caution against naïve investment in the apparent 'truths' offered by the movies. By repeatedly qualifying the dramatic urban imaginary fostered by popular Hindi film, Mehta's polyphonic narrative encourages a more scrupulous analysis of the sidelined people and places that are too often passively consumed as disposable entertainment.

Urban waste from 'the ideal position': Mulk Raj Anand's Untouchable

While a number of critics have identified the cinematic intertextuality of *Maximum City*, drawing attention to Mehta's literary depiction of Bombay's visually arresting urban dramas, few have considered his place within a tradition of Indian urban writing that similarly encourages more careful analysis of the mutually constitutive real and imaginary geographies that shape the margins of India's cities. In writing about his childhood home after spending twenty-one years in the USA and Europe, Mehta joins a number of literary predecessors to offer a hybrid migrant's perspective on India's cities. Mulk Raj Anand initiates this tradition with *Untouchable* (1935), a short novel that offers an important critique of the social inequalities perpetuated by the Indian caste hierarchy. This longstanding mode of social stratification divides people into thousands of endogamous hereditary groups. Under this system, evident filth – animal dung, human excrement, household trash – is collectively ignored by the upper castes who defer responsibility for its management and removal to the Dalits or 'untouchables' who perform an essential, but unacknowledged social function. While significant measures have been taken since the adoption of the Indian Constitution in 1950 to address the discriminatory treatment of Dalits, with many scholars arguing that British colonialism is responsible for not only manipulating, but in some cases manufacturing social division, the historical allocation of material waste management to symbolically 'unclean' social groups provides a formative context for the contradictory dynamics of visibility that pervade contemporary Bombay.[7] To be an untouchable, as Anand reveals, is precisely to inhabit the jarring simultaneity of material exposure and social invisibility that affects contemporary slum-dwellers.

As a Cambridge-educated Indian writer, immersed in early twentieth-century literary culture, Anand offers a useful bifocal perspective on the caste system. E. M. Forster's preface to *Untouchable*, itself a testament to Anand's affiliation with the Bloomsbury Group, notes that the novel:

> could only have been written by an Indian, and by an Indian who observed from the outside. No European, however sympathetic, could have created the character of Bakha, because he would not have known enough about his troubles. And no Untouchable could have written the book, because he would have been involved in indignation and self-pity. . . . Mr. Anand stands in the ideal position. . . . He has just the right mixture of insight and detachment.
>
> (Anand 2009 [1935], pp. vi–vii)

Here, Forster repeats the condescending definition of Dalits as something other than Indians or Europeans; they are identified by their untouchability and implied emotional immaturity. Ironically, despite his own recourse to polarising social categories, Forster identifies in Anand's unique viewpoint a critical insight that his own commentary lacks. From his liminal perspective, Anand can persuasively critique the rigidity of a caste system that disavows the shifting social determination of both literal and figurative waste, even as modern urbanisation drastically reshapes these categories. Moreover, as Ben Baer notes, Anand's novel expresses 'the desire to carry the periphery to the metropolis so as to inscribe and make visible the unknown, excremental abjection of the colonial margin in the aesthetic heart of the center' (Baer 2009, p. 577). While both the upper castes and colonial classes invoke Indian urban waste as the marker of social boundaries, Anand's engagement with Dalit experience reveals these categories to be inadequate to both local reality and the broader transnational context through which they were produced.

The pernicious psychological effects of the caste system are focalised through the eighteen-year-old Bakha, a sweeper and latrine cleaner whose innate pride, curiosity and sense of social injustice leads him to question the actual and imagined boundaries that maintain his social marginalisation: 'He was a sweeper, he knew, but he could not consciously accept that fact' (Anand 2009, p. 39). Both figuratively and physically marginalised, Bakha lives with his family in the outcastes' colony adjacent to the small city of Bulashah. However, it is when he ventures into the urban centre to sweep streets on behalf of his unwell father that his social standing is most destabilised. Here, in the city, the boundaries that maintain Bakha's lowly untouchable status are thrown into confusion by the sensual and visual distractions of urban modernity. Treating himself to some confectionery from a corner store, the pleasure of Bakha's physical indulgence combines with his voracious visual consumption of the city's commercial spectacles to inspire his intellectual and social ambition:

> It was wonderful to walk along like that, munching and looking at all the sights. The big signboards advertising the names of Indian merchants, lawyers and medical men, their degrees and professions all in broad, huge blocks of letters, stared down at him from the upper stories of the shops. He wished he could read all the luridly painted boards. . . . Then his gaze was drawn to a figure sitting in a window. He stared at her absorbed and unselfconscious.
>
> (Anand 2009, p. 46)

His freedom is short-lived however; distracted by the unnamed woman – Anand hints very lightly here at the sexual awakening attendant on Bakha's entrance into the city – he walks into a higher-caste man, thereby violating the symbolic distance that is supposed to be kept between untouchables and those higher up the social ladder. The man responds with violent physical and verbal abuse, repeatedly casting Bakha in bestial, non-human terms: 'Keep to the side of the road, you low-caste vermin! . . . Why don't you call, you swine, and announce your approach! . . . Dirty dog! Son of a bitch! The offspring of a pig!' (Anand 2009, pp. 46–47). This traumatic accident renews Bakha's painful consciousness of his untouchable status: 'Like a ray of light shooting through the darkness, the recognition of his position, the significance of his lot dawned upon him' (Anand 2009, p. 52). His eyes drawn to the pleasures and possibilities of city living, Bakha literally fails to look where he is going and, in doing so, he forces an inter-caste contact which shatters his socially constructed invisibility. By touching someone of higher social standing, he destabilises the collective ignorance of waste maintained by his social 'superiors'. The 'defiled' man is compelled to recognise Bakha and the filth in which he works; thus, he is compelled to acknowledge his own role in the production of that filth in both its physical and figurative aspects. By portraying this eruption of social tension from Bakha's perspective, Anand artfully highlights the constructedness of the caste system, suggesting the need for an alternative, more equitable mode of social organisation that is in keeping with the new possibilities engendered by urban modernity. This crucial episode reveals not only the pressure that urbanisation places on the caste system's allocation of filth and cleanliness to different groups, but also the particular visual register in which this occurs.

V. S. Naipaul's Indian vision

In keeping with Anand's appraisal, Trinidadian author V. S. Naipaul expresses distaste for the 'dangerous, decayed pragmatism' of the caste system in his Indian trilogy: *An Area of Darkness* (1964), *India: A Wounded Civilization* (1976) and *India: A Million Mutinies Now* (1990) (1964, p. 85). Yet if, as Baer suggests, Anand draws on his bifurcated perspective to problematise the gap between metropole and colony, the similarly European-educated Naipaul asserts what some critics have argued is a deeply imperial viewpoint throughout his Indian travel writing. Instead of writing *to* a metropolitan audience, he writes *for* them, in what Rob Nixon calls 'a language that resonates with traditions of discursive power that assert the visual and political ascendancy of metropolitan knowers over the peripheral and underrepresented known' (Nixon 1992, p. 80). Indeed, Naipaul himself openly commends the 'Western way of perceiving' as superior to Indians' 'shallow perception' of reality (Naipaul 1976, pp. 108, 119). He laments the supposed inability of Indians to analyse their situation objectively and, in doing so, ameliorate not only the material deprivations of their rapidly modernising country, but also what he sees as its postcolonial intellectual poverty. His critique of independent India's confused, imitative culture – ironic in the context of his own imperial mimicry – meets with a visceral disgust for the messy reality of the country, especially Bombay, to produce what many have read as a darkly pessimistic account of his ancestral home.

While *India: A Wounded Civilisation* extends Naipaul's critique of the Indian 'defect of vision' (Naipaul 1976, p. 101), the third volume in his Indian trilogy, *India: A Million Mutinies Now*, has been read as a 'kinder, gentler' account of the country (Nixon 1992, p. 159). Although the opening section, 'Bombay Theatre', rehearses some of the same tropes of urban filth that characterised his earlier portraits of the city – shortly after arrival, for example, he notes buildings weathered by both 'excessive sun, excessive rain, excessive heat' and 'human grime' (Naipaul 1990, p. 1) – he does acknowledge the 'old neurosis' revived by his earlier visits to India. As he explains, confronting the poverty that his emigrant grandparents had sought to leave behind in moving to Trinidad destabilised the dignified collective identity they had so carefully cultivated in diaspora. His first trip to India thus profoundly undermined the idealised cultural heritage through which the young Naipaul had identified himself. Yet despite acknowledging the deeply personal anxiety that previously filtered his view of India, Naipaul can still not bring himself to fully engage with the most abject of urban existence. His account of Bombay does document the political rise of the Dalits through interviews with representatives of their caste, which take place in their modest homes. Yet for all its clear-sightedness, his urban portrait retains a significant blindspot with regard to the least appealing areas of the city. Glimpsed in passing, the infamous slum of Dharavi appears to Naipaul 'so sudden, so obvious, so overwhelming, it was as though it was something staged, something on a film set, with people acting out their roles as slum-dwellers' (Naipaul 1990, p. 58). A subsequent drive past the slum makes a similarly surreal impression on the author; seen from the taxi, 'Dharavi looked artificial, unnecessary even in Bombay' (Naipaul 1990, p. 59). While his choice of cinematic metaphor is both understandable given the shocking extent of Dharavi, where an estimated 1 million people live in an area of approximately 1.7 square kilometres, and apt in the context of Bombay's thriving film industry, it also speaks to Naipaul's reflexive retreat from the urban waste which he has earlier criticised Indians for ignoring. Elegant description here substitutes for full engagement with Bombay's margins. By comparing Dharavi to a film set viewed through the 'screen' of his car window, Naipaul self-defensively restores the cinematic distance that is shattered for Anu in the police station. Whereas Mehta calls attention to the limits of filmic mimesis to encourage greater sensitivity to the challenging reality that popular films often mask, Naipaul invokes cinematic artificiality to avoid precisely such engagement. Instead, the Bombay that emerges from his account in *India: A Wounded Civilisation* consists in finely rendered portraits of individually aspiring characters in various carefully described interiors, the interconnections between which are not fully elaborated. Bombay's urban waste remains sidelined as the city's dirty secret that Naipaul is unwilling to reveal.

Seeing 'the other Bombay'

Having seen, and smelled, Dharavi from a distance, Naipaul expresses his 'relief to leave that behind, and to get out into the other Bombay, the Bombay one knew and had spent so much time getting used to, the Bombay of paved roads and buses and people in lightweight clothes' (Naipaul 1990, p. 59). The urban divide he references here is perhaps not entirely lacking in self-deprecating irony, but it is also characteristic of Naipaul's inability to reconcile his singular expectations of India with the unpredictable, divided reality he confronts.

Suketu Mehta also invokes Bombay's duality in *Maximum City*, announcing his intention to explore 'the other Bombay' in a phrase that directly recalls Naipaul. However, his account of the city quickly reveals its interconnectedness, the denial of which produces the slums' paradoxical visibility as those who do not wish to confront the evident social inequalities before their eyes must engage in mental and moral contortions in order to repress what they have witnessed. On his return to India after living abroad for twenty-one years, Mehta literally confronts the space between reality and recall. He is struck by Bombay's rapidly changing physical and political landscape, exemplified by the inescapable encroachment of informal housing throughout the city, such as the shantytowns that have sprung up on the beach where he fondly remembers playing as a child. Unlike Naipaul who, throughout *Area of Darkness*, cultivates a personal distance from what he repeatedly refers to as the 'obvious' poverty of India, Mehta decides to investigate the marginal, unplanned and informally regulated city of slums and criminal activity that is gradually encroaching on the insular world of the Bombay elite to which he belongs as an educated, cosmopolitan writer. Whereas his predecessor finds that his 'eye had changed' during the course of his travels around the country, enabling him to 'separate [him]self from what [he] saw, to separate the pleasant from the unpleasant', Mehta develops a more nuanced view of India that troubles the binary terms of Naipaul's Indian vision, which so readily distances spectator from subject, clean from dirty (Naipaul 1964, p. 48). As both a material problem and a symbolic category, Mehta finds waste in contemporary Bombay to be multidirectional and multiscalar.

Somewhat ironically, Mehta comes to acknowledge the interconnectedness of the city through what might be described as a Naipaulian frustration with the city's inadequate municipal services. Early in the text, he expresses in humorously exaggerated terms the perilously unsanitary water supply to which his not inexpensive rented apartment is connected:

> The food and the water in Bombay, India's most modern city, are contaminated with shit. Amebic dysentery is transferred through shit. We have been feeding our son shit. It could have come in the mango we gave him; it could have been in the pool we took him swimming in. it could have come from the taps in our own home, since the drainage pipes in Bombay, laid out during British times, leak into the freshwater pipes that run right alongside. There is no defense possible. Everything is recycled in this filthy country, which poisons its children, raising them on a diet of its own shit.
>
> (Mehta 2004, p. 28)

If, as Dipesh Chakrabarty notes, the domestic space is 'an inside produced by symbolic enclosure for the purpose of protection' (Chakrabarty 2002, p. 71), Mehta's emphatic sentence-end repetition of 'shit' conveys his horrified realisation of the inadequacy of this construct in the face of Bombay's systemic sanitation problems. To live in the city is to acquire an unwanted, cellular intimacy with its many residents. With 'no defense possible', he feels besieged by the filth of the city. By noting the colonial origins of the city's profoundly inadequate sewage and water supply, Mehta shifts characteristically from the microscopic to the macroscopic, reminding the reader that Bombay's most immediate and intensely personally experienced problems have a long history and a global context. He offers little indication of any solution to the problem, noting that fecal matter is embedded in the very structure of the numerous new buildings that are proliferating in the city: 'The sand used in the concrete comes from the creeks around Bombay, which contains salt, silt, and shit, so new buildings look weather-beaten, moth-eaten' (Mehta 2004, p. 120). Shit circulates through both the bodies and the buildings of Bombay, connecting humans and their housing through an inescapable cycle of waste that the wealthy wish to ignore and the poor cannot.

If Bombay's infrastructural weaknesses expose the fault-lines in the carefully constructed social boundaries between the city's wealthy and working classes, the riots which break out in late 1992, shortly prior to Mehta's return, explicitly dismantle them. Catalysed by the destruction of a northern mosque by Hindu fundamentalists, violence exploded in Bombay with the leader of the Shiv Sena Hindu nationalist party Bal Thackeray inciting the city's disaffected Hindu poor to systematically eliminate the city's Muslim residents, homes and businesses. In

March 1993, a series of bombs planted by the Muslim underworld were detonated throughout the city, in apparent revenge for the losses inflicted upon Bombay's Muslim community. Many consider the riots to be 'a milestone in the psychic life of the city', marking the transition from the secular cosmopolitanism of Bombay to the ethnic chauvinism of Mumbai (Mehta 2004, p. 56). With the eruption of street violence and the subsequent bombings of prominent public places, Bombay's most marginalised urban residents became dramatically visible to the city's richer inhabitants. An educated man from a family of Gujurati diamond merchants, Mehta enjoys relative wealth and privilege. Yet while he is sufficiently self-aware to recognise the tenuousness of his social status after the riots, he notes that many of his peers share 'a sense that the barbarians have been let into the city gates and are sleeping on the footpaths outside their palaces. There is resentment that Bombay has to deal with the country's detritus' (Mehta, 2004, p. 76). What these detached upper classes fail to account for in their ongoing dismissal of the supposed human waste that accumulates in the city is the very real political power that the urban poor wield in local and national elections by their sheer numbers alone. Although Naipaul does not fully engage with the reality of Bombay's slums, he correctly identifies them as key sites of this sociopolitical shift. Dharavi intimidates not only because it is a reminder of the struggles that Naipaul and others serendipitously evade, but also because it is 'a vote-bank, a hate-bank, something to be drawn upon by many people' (Naipaul 1990, p. 60). As Mehta bluntly puts it, 'In India, the poor vote' (Mehta 2004, p. 68). Surplus they may be to the global capitalist system in which Bombay is embedded, but the city's slum- and street-dwellers are not afraid to reclaim their outcaste status by 'getting [their] feet dirty in politics' (Mehta 2004, p. 75). Although, Mehta notes, 'the monster came out of the slums' during the riots, he resists the dismissive sensationalism this metaphor implies, choosing instead to investigate the motives of those who got involved (Mehta 2004, p. 56).

Mehta's re-vision of the city draws on lived experience, not removed observation. As a result, he is able to perceive the kaleidoscopic reality of Bombay, whose endless shifts refuse simple typologies. In the course of his investigation into the causes of the riots, Mehta inhales the 'stench' of the slums in which the majority of the Shiv Sena supporters live (Mehta 2004, p. 53); he treads the 'pitch-dark alley[s]' where illegal settlements are being constructed (Mehta 2004, p. 77); he drinks 'rich thick buffalo milk' in the home of Amol, a tapori ('street punk') who incites violence on behalf of the Shiv Sena (Mehta 2004, p. 85); and he hears first-hand accounts of the riots from a range of victims, instigators and witnesses. Unlike Naipaul, he ventures into precisely those parts of the city that are habitually ignored by class and caste outsiders. In doing so, he reveals the rich multipurpose nature of putatively wasted areas. One day, for example, Raghav, a private taxi operator and unofficial Shiv Sena troublemaker, takes Mehta to what he describes as

> A very large open patch of ground by the train sheds, a phantasmagoric scene with a vast garbage dump on one side with groups of people hacking at the ground with picks, a crowd of boys playing cricket, sewers running at our

feet, train tracks and bogies in sheds in the middle distance, and a series of concrete tower blocks in the background.

(Mehta 2004, p. 44)

As Mehta explains, this is the no-man's land between Hindu- and Muslim-populated slums. With characteristic detail, Mehta's description populates a tract of land that appears as vacant industrial storage space on official city plans. He later meets with an architect who identifies these spaces as ideal for redevelopment. What such conceptions of Bombay's wasteland fail to accommodate are the multiple uses to which such land has already been put – this is a recreational space for the slum kids who have nowhere else to play and a site of potential income for the scavengers who sort through the trash. Most significantly, this is a site of trauma and oral history, as Mehta's discussion with Raghav reveals. During the riots, Raghav and some Sena loyalists beat and burned two Muslims on this site. He further explains that, for ten days, 'the police wouldn't take the bodies away, because the Jogeshwari police said it was in the Goreagon police's jurisdiction, and the Goreagon police said it was the railway police's jurisdiction' (Mehta 2004, p. 45). In their refusal to dispose of the victims' bodies, the various civic authorities expose their petty and damaging adherence to a rigid conception of urban space that cannot accommodate the 'phantasmagoric' region that Mehta exposes. Like the Shiv Sena, who are only concerned with policing the ethnic content of their slum, the police attribute a damaging singular value to this urban edgeland, which results in further social and environmental degradation. By revealing the many valences of this urban space, Mehta brings to light marginal areas that others deliberately ignore. In doing so, he vividly illustrates the inimical capacity of rigid urban borders to produce new forms of waste.

Mehta's ethnographic authority

Despite his efforts to witness the slums from the 'inside', Mehta is still a privileged outsider who relies on a series of 'native informants' like Raghav to provide him with 'a tour of the battlegrounds' (Mehta 2004, p. 40). Indeed, in his desire to find out more about the causes of the riots – 'I wanted to speak to the rioters themselves. . . . I wanted to find out how the business of rioting is actually planned and carried out' (Mehta 2004, pp. 40–41) – Mehta repeatedly expresses an extractive desire for knowledge about the slums, which casts him in the role of a colonial fact-finder surveying unfamiliar terrain. However, to dismiss his project as imperialist would be unfair to the careful manner in which he draws attention to his own limited vision on a number of occasions. Concerned with revealing differing perceptions of Bombay's margins, Mehta rejects the role of omniscient narrator. This is evident, for example, in his interactions with Sunil, a Shiv Sena agitator who calmly describes setting fire to a Muslim bread-seller whom he used to pass on a daily basis before the riots began. When Mehta asks 'what does a man look like when he's on fire?' Sunil responds, '*You* couldn't bear to see it' (Mehta 2004, p. 39). The two men have entirely different perspectives on Bombay. They see the city differently and, in doing so, produce multiple lived versions of that city. Indeed, when Sunil enters the apartment in which one of their interviews takes place, he 'immediately check[s] out the strategic value of its location, its entrances and exits' (Mehta 2004, p. 42). Differently attuned to urban space, Mehta is reliant on Sunil's first-hand knowledge to produce a convincing account of the riots. Using a form of free indirect discourse, he thus transcribes large portions of Sunil's oral testimony, which explain how and why this disaffected young man committed murder.

As James Clifford notes in his highly influential essay 'On Ethnographic Authority', even such commendable attempts at diffusing the false claims of experiential omniscience are tempered by the very fact of ethnographic authorship. As he explains, 'While ethnographies cast as encounters between two individuals may successfully dramatise the intersubjective give-and-take of fieldwork and introduce a counterpoint of authorial voices, they remain *representations* of dialogue' (Clifford 1988, p. 43). Although directly quoting Sunil's words marks an attempt on Mehta's part to distribute the authority of his account, he nonetheless remains the arranger of those words – he chooses what to include and omit, and, most importantly, he translates Sunil's words from Hindi into English, lending him a textual eloquence that belies writerly detachment.

Interestingly, Mehta is willing to withhold such eloquence from other informants, such as Bal Thackeray, the notorious Shiv Sena leader whom Mehta, like many others, holds largely responsible for the 1992–93 riots. When the two finally meet in Thackeray's heavily guarded urban compound, they converse in what Mehta describes as 'fractured English' (Mehta 2004, p. 97). Whereas the direct quotations that Mehta includes from Sunil encourage, if not sympathy for, then at least some understanding of his participation in the riots, the inclusion of extended oral testimony from Thackeray serves to make a mockery of the Shiv Sena leader. Mehta reproduces long pronouncements from Thackeray which fully display his struggling command of English, their overall effect being to diminish his intelligence. Mehta admits to 'entertain[ing] the suspicion that he is not all there' when Thackeray embarks on soliloquies concerning what he perceives as Bombay's infrastructural and immigration problems (Mehta 2004, p. 101). Mehta refuses dialogue in this section of *Maximum City*, withholding the narrative compromise that he allowed Sunil and his other less powerful interlocutors. While Thackeray personifies a dangerously essentialist cultural ideology that imagines a renamed Mumbai as the capital of an exclusively Hindu nation, it is ironically by adopting the party's policy of non-negotiation that Mehta is able to undermine him. Lacking sympathetic translation, Thackeray is ultimately condemned by his own, rambling words.

Mehta's encounters with Sunil, Satish and Thackeray derive tension from his self-dramatisation. Whether wandering through the slums, meeting with gangsters in confined hotel rooms, or entering the politician's heavily guarded compound, Mehta conveys to the reader that he is himself in some kind of jeopardy. This authorial approach contrasts with that of Katherine Boo, whose *Behind the Beautiful Forevers* (Boo 2012) is an elegantly written and thoroughly researched narrative of life in Annawadi, 'a single, unexceptional slum' squeezed between the luxury hotels adjacent to Mumbai international airport, the inhabitants of which make their living from scavenging and sorting waste (Boo 2012, p. 249). A renowned journalist like Mehta, Boo's text expands her previous work on poor American communities in an attempt to illuminate the 'infrastructure of opportunity' in Indian society (Boo 2012, p. 247). Whereas Mehta's own experiences lend drama to his text, suggesting another subjective Bombay to be taken into consideration, Boo effaces herself from her narrative, reserving self-disclosure for her concluding author's note in which she explains her careful attempts to compensate for the limitations of her outsider's perspective 'by time spent, attention paid, documentation secured, accounts cross-checked' (Boo 2012, p. 249). Although she asserts that she 'was mindful of the risk of overinterpretation,' her decision to withhold her own experiences from the text in favour of assuming a third-person narrative voice at times lends her narrative the tone of mythic omniscience (Boo 2012, p. 250). It begins, for example, in *medias res* as she relates the dramatic self-immolation of a disabled Annawadi resident: 'Midnight was closing in, the one-legged woman was grievously burned and the Mumbai police were coming for Abdul and his father' (Boo 2012, p. ix). This effective opening sentence grabs the reader's attention and lends her slum-dwelling subjects an air of epic significance. Her narrative style effectively makes of her Annawadi subjects a universal drama. This broad relevance is in keeping with Boo's stated interest in the patterns of 'profound and juxtaposed inequality' that form 'the signature fact of so many modern cities' (Boo 2012, p. 248). While Mehta's personal narrative occasionally touches on narcissism, Boo's conscious distancing from her subjects avoids this pitfall of an otherwise insightful self-reflexivity. However, the fluency with which she reconstructs the stories of the slum-dwellers by retrospectively paraphrasing their thoughts and words risks obscuring the actual difficulty that these marginal urban residents face in their daily struggles to be seen, heard and recognised on their own terms.

Gendered city

Boo not only provides a useful stylistic complement to Mehta's narrative, she also offers a female counterpoint to his male perspective on the city. One of few women to write so extensively about slum existence, her sympathetic attention to the struggles faced by female slum-dwellers, specifically documenting a spate of female suicides motivated by gendered inequities in slum society, provides an informative contrast with the male-dominated world of criminality and violence that Mehta largely engages. If Mehta exhibits a self-reflexiveness that Boo chooses not to articulate in the body of her text, this is limited by his blindness to the gendering of his own gaze.

Mehta's well-intentioned attempts to diffuse his own ethnographic authority are compromised by his interactions with an exotic dancer who performs under the name of Monalisa. Whereas Mehta tries to cultivate neutral detachment when dealing with the male Shiv Sena agitators, he casts an undeniably masculine gaze on Monalisa, which compromises the objectivity of his narrative. The uneven power differential exposed by the interactions between male author and female subject further highlight Monalisa's limited ability to rescript her own status as urban 'waste'. Unlike the male Shiv Sena agitators who enjoy a degree of autonomy through their strategic embrace of their 'wasted' status, the urban freedoms of Monalisa and other marginalised women like her are more tightly circumscribed by gender expectations. Indeed, Bombay's contradictory dynamics of visibility are dramatically inscribed on her gendered body, which is at once exposed to objectifying gazes and concealed from the respectability of public view.

Monalisa is a dancer in what is known in Bombay as 'the bar line' – the collective term for the many 'beer bars or ladies' bars or dance bars' throughout the city in which 'fully clothed young girls dance on an extravagantly decorated stage to recorded Hindi film music, and men come to watch, shower money over their heads, and fall in love' (Mehta 2004, p. 265). If Bombay is a site for the acquisition and consolidation of 'power', as the title of Part I suggests, Part II reveals the city to be a complementary locus of 'pleasure' where other desires may be satiated. However, Mehta's separation of these two aspects of Bombay is quickly exposed as a falsehood since his narration of the city's indulgent and luxurious nightlife repeatedly points to the manner in which soliciting pleasure always entails the negotiation of power differentials for all involved.

Mehta claims curiosity as the initial catalyst for his investigation into the bar line: 'I started going to the beer bars because I was puzzled. I couldn't figure out why men would want to spend colossal amounts of money there' (Mehta 2004, p. 269). Before long, however, this bemused outsider not only understands the bar's male clientele, he epitomises them in his newfound role as 'best guide' to the bar scene, enjoying VIP treatment at Sapphire, the bar where he meets Monalisa (Mehta 2004, p. 338). He identifies a marginal form of urban democracy in the bar, observing that:

> This is the one place where the classes meet, where the only thing important is the color of your money. Because it's not just the mechanics and the taporis; it's also the rich traders and merchants of South Bombay, who are surrounded by men during the day and by their fat wives in the evening. . . . The moment the customer walks in, he's the star in his own custom-made Hindi movie song. No matter how old or ugly or fat he is, for the two hours he's in the bar he's a movie star, he's Shahrukh Khan.
>
> (Mehta 2004, p. 271)

This idealised description of the bar's escapist promise obscures the uneven access to power enjoyed by the male customers and the dancers who meet in this apparently democratic night-time space. Whereas the men who can afford to frequent the bar are at liberty to gaze upon 'beautiful young girls, young enough to be their daughters', the women are have limited control over their self-presentation (Mehta 2004, p. 271). Bollywood scripts are again invoked to mediate and mask the degradation of Bombay's margins. In living out their cinematic fantasies, the bar line's male customers perpetuate the social exclusion and financial dependency of those women who fulfill them. As Mehta later explains, even those dancers who are justifiably proud of their choreography, performance skills and physical beauty are nevertheless demeaned by their occupation.

Mehta's description of first being taken to the bar further suggests the limited control that the bar line dancers have over their working conditions. He is inducted into the bar scene by a male friend who escorts him to a 'completely dark alley' in Worli (Mehta 2004, p. 269). When they reach the Carnival Bar that is supposed to be closed, various men appear from the shadows to park their car and usher them into what turns out to be a bar 'ablaze with light and music and flowing with liquor and filled with people at 3 a.m.' (Mehta 2004, p. 270). Although Mehta suggests that the women who dance in the bars may enjoy the attention they receive to some extent, the power they wield on stage is tightly circumscribed by the male-dominated urban world in which they live. Men are the gatekeepers, managers and consumers of their dances – without them, the women would be unable to make money. Their dependent situation epitomises the contradictory dynamics of visibility that run through Bombay; although the women are highly visible within the bars – indeed, they make money based on how skilfully they reveal their physical beauty – the bars themselves are a covert space where men indulge their fantasies of being movie stars.

While, as Altaf Tyrewala suggests, Mehta is 'astute' in his account of bar line life, portraying as he does both its sensational and mundane realities, his narrative also exerts additional authority over the bodies of the dancers, especially that of Monalisa (Tyrewala 2012, p. 16).[8] Just as the men who manage Sapphire control when, how and by whom Monalisa is seen, Mehta's words similarly manipulate the way in which she is presented to a new, literary audience. Although his account is both attentive and sympathetic to the details of Monalisa's troubled life, its effect is ultimately to offer up an additional performance for his readers, compounding Monalisa's struggles to achieve a coherent identity.

Like the bar clientele who enjoy, for a short time, the illusion of 'star[ring] in [their] own custom-made Hindi movie song', Mehta's interactions with Monalisa blur fantasy and reality from the outset (Mehta 2004, p. 271). He recalls that when he first saw her, 'all the other girls blurred and faded, as in a movie when the heroine suddenly comes into sharp focus as she's walking in a crowd of people in the street' (Mehta 2004, p. 273). By casting himself in the role of a hopelessly attracted and naïve suitor, who 'has to make a conscious effort to keep [his] hand from trembling' when he shows her into his apartment for an interview, Mehta lends Monalisa a power over him that she does not in reality have (Mehta 2004, p. 283). His attraction to her is surely genuine – she is young, beautiful and expert in the art of seduction. However, it is Monalisa who reveals herself completely to Mehta, and not vice versa. She exposes herself physically and psychologically to him, confiding secrets, aspirations and disappointments, whereas he withholds information from her during their intimate liaisons, including the fact that he is married and has two children. Mehta professes paternal responsibility for doing so, explaining that Monalisa 'is of the shadow world; I keep my family insulated from such people' (Mehta 2004, p. 295). However, he does not extend such protectiveness to his equally vulnerable female interlocutor. The metaphor he uses here is telling, suggestive as it is of the way in which Monalisa's visibility is always partial and fleeting. Although she is inseparable from Mehta's world, she does not control her connection to it. Like the men who control access to the bars, Mehta also acts as a gatekeeper, maintaining distance between his family and Monalisa, barring access to his wife.

Mehta's first-hand accounts of Monalisa's on-stage performances make his authorial staging of her body explicit. Although his descriptions speak to the dancer's powerful seductive charms, the narrative also reproduces the uneven power dynamics of the bar, offering her up for visual consumption by an unseen audience. The reader gazes upon Monalisa from Mehta's own guarded perspective, voyeuristically watching as 'this girl . . . turn[s] her back to the audience, bend[s] forward, and slowly rotat[es] her buttocks' (Mehta 2004, p. 273). Monalisa enacts the fantasy of sexual availability while her onlookers remain anonymous in the privacy of the bar. Although Mehta elsewhere expresses a seemingly genuine interest in her story, his gaze, like that of the other male spectators, objectifies Monalisa. His tendency to catalogue her physical features as he watches her reduces her to an assortment of attractive body parts: 'She had big bee-stung lips, a high neck, large eyes, and a snub nose' (Mehta 2004, p. 273). Taken in by her act, he essentialises what he perceives as her primal energy: 'This young Gujurati girl becomes, on the dance floor, an animal with not enough space to move' (Mehta 2004, p. 281). Interestingly, even as he 'get[s] to see [Monalisa's] nice side' over time, Mehta tells himself not to 'forget her core, which is based on sex, on lust' (Mehta 2004, p. 303). In his eyes, her sexuality is her primary identification.

Mehta tempers the intrusiveness of his gaze by insisting that Monalisa 'likes being looked at, likes being noticed' (Mehta 2004, p. 281). Indeed, she expresses legitimate pride in her ability to command attention. However, Mehta also mistakes Monalisa's learned survival instinct for her public self-confidence. Monalisa's success in the bar line, which she calls 'a world of lies', depends on her ability to project an assured sexuality even when she feels tired, vulnerable or uncertain (Mehta 2004, p. 293). Living as she does 'two lives' – 'one is her life in the bar and the time she spends with her customers. Then there is the other life: her time in the discos, watching TV, sleeping all day' – Monalisa is adept at role-play, expert in fulfilling the fantasies of the men who come to watch her and concealing her professional persona in public (Mehta 2004, p. 293). Her personal life similarly demands the suppression of her emotions. Having experienced multiple family traumas, Monalisa has learned how to keep her feelings in check: 'If Monalisa were to allow herself to cry every time she felt the weight of pain or heightened emotion, she would be all dried out from the crying' (Mehta 2004, p. 313). Monalisa's continual subjection to intense scrutiny demands great composure on her part, an effort that Mehta overlooks in his celebration of her life lived 'on the extreme of spectacle': 'The attraction, the immense relief, of total breakdown, a renunciation of order in one's life, of all the effort required to keep it together!' (Mehta 2004, p. 538). Where he sees 'freedom' and 'a life unencumbered by minutiae', he misses the strain under which Monalisa's bifurcated existence places her.

Mehta is seemingly unaware of the way in which his own narrative actually makes Monalisa vulnerable in a way that even bar dancing does not. Attractive as her body is, it also disappoints him, as is revealed when he arranges for his fashion photographer friend Rustom to take some test shots of her. Again, Mehta enumerates her physical features, but this time it is the flaws that appear in the brightly lit studio: 'Under her black velvet top, I notice for the first time that she has a small paunch, a belly that has popped out. Her smile is crooked; her lips curl up at the extreme left of her mouth' (Mehta p. 296). Although relating the studio episode signifies a self-aware and humorous puncturing of Mehta's own fixation with Monalisa, it also destroys the carefully cultivated bar line persona that enables her to earn a living in challenging circumstances. There is a cruelty to Mehta's dismantling of the alluring image that Monalisa seeks to maintain. In this instance, again, the power dynamic implied by their different gazes is made evident. What, for Mehta, is a moment of knowing narrative exposition, is for Monalisa, a moment of extreme vulnerability. His critical gaze wrests control of her image from her.

Especially troubling are 'the marks all down [Monalisa's] arms [that] can be seen clearly under the glare of the powerful studio lights' (Mehta 2004, p. 286). Mehta has earlier observed that her 'wrist is scarred and pitted like a dirt road' – the result of multiple suicide attempts. Monalisa habitually cuts herself during episodes of high emotional stress, literally inscribing the pain of being a young, marginalised woman in Bombay onto her body. Her potentially fatal behaviour is sadly not unique to her situation. While at Rustom's studio, Monalisa encounters Marika, 'the hottest model in the country', and immediately notices her similar scars, which both the male writer and male photographer present have overlooked (Mehta 2004, p. 282). The extent of hidden female suffering in Bombay suddenly visible to him, Mehta realises that:

> There must be a citywide sorority of these women who've slit their wrists and survived, who recognise one another automatically. A sisterhood of the slashed. The top model in India and the top bar dancer in Bombay have this in common: Their arms are marked with their anguish, like gang tattoos.
>
> (Mehta 2004, p. 283)

Again, Bombay's realms of 'pleasure' and 'power' are conflated. Whereas Mehta openly professes his enjoyment of the 'beery fraternity' that Sapphire fosters among its diverse male customers, this scarred 'sisterhood' levels the city's hierarchy in a far more troubling manner (Mehta 2004, p. 313). With painful irony, Monalisa and her female peers resort to self-harm in an attempt to assert control of their own bodies and identities. Mehta's casual equation of the women's scars to 'gang tattoos' reinforces the fact that many of Bombay's marginalised men strategically embrace their 'wasted' status to wield relative power as gangsters, criminals, dealers and pimps. Women, however, experience heightened vulnerability. Even when they are as successful as the model Marika, they struggle to achieve their desired self-presentation, valued as they are by many for their sexuality alone.

For Mehta, however, narrative desire ultimately sublimates sexual desire for Monalisa, as the final paragraph of this section reveals:

> At some point the Monalisa I'm writing in these pages will become more real, more alluring, than the Monalisa that is flesh and blood. One more ulloo, Monalisa will think. But imagine her surprise when she sees that what I am adoring, what I am obsessed with, is a girl beyond herself, larger than herself in the mirror beyond her, and it is her that I'm blowing all my money on, it is her that I'm getting to spin and twirl under the confetti of my words. The more I write, the faster my Monalisa dances.
>
> (Mehta 2004, p. 314)

Ironically, this denouement both exposes and conceals Monalisa. By suggesting that she is naïve to assume that he has fallen for her, Mehta retains the upper hand in their relationship. His assertion of authorial control undermines her attempts to be seen and heard on her own terms. In Mehta's narrative, as in the bar, on the streets, and in the studio, the 'real' Monalisa is relegated to the shadows while her body becomes a site for the projection of male fantasies.

Conclusion

Mehta's narrative staging of Monalisa's marginal body reveals the precarious task undertaken by authors of the contemporary postcolonial city. On the one hand, in keeping with his other thoughtful portrayals of Bombay's strugglers, outliers and transients in *Maximum City*, Mehta's account of Monalisa's marginal bar line life directs attention to a hidden dimension of urban reality, the necessarily covert nature of which inhibits legitimate claims to full social recognition and rights on the part of such women. However, the gendered gaze through which Mehta perceives Monalisa compromises his critique of the contradictory dynamics of visibility to which such female urban residents are subject, instead intensifying their capture within a disjunctive matrix of exposure and invisibility over which they have limited control.

Although problematic, the inherent ambiguity of Mehta's narrative highlights his contention that Bombay existence emerges from the dynamic interplay between urban reality and urban fantasy. As his critical engagement with popular Hindi film suggests, any account of Bombay, whether literary or cinematic, should not be taken at face value. Following in the migrant literary tradition of both Mulk Raj Anand and V. S. Naipaul, both of whom demonstrate, albeit in contrasting fashion, the ways in which Indian urban modernity disrupts the visual dynamics that keep city margins in place, Mehta offers a compelling critique of how, why and by whom Bombay's most degraded places and their dejected occupants come to be seen as urban 'waste'. Through careful description, qualified ethnography and personal reflection, he thus demands their immediate revaluation both as legitimate actors within the modern Indian state and as disenfranchised participants within an uneven global system that reinforces their continuing marginalisation.

Notes

1 Although Bombay was officially renamed Mumbai in November 1995, I retain the use of the former throughout this chapter in keeping with the preference of Suketu Mehta, whose nonfictional account of the city is analysed here. He, like many others, sees the renaming of the city as a political move on behalf of the then newly elected Shiv Sena nationalist party to strengthen Hindu identity throughout the Maharashta region and the country as a whole. If the Anglicised name 'Bombay' bespeaks an unwelcome European colonial legacy, 'Mumbai' is, in Mehta's view, allied with an equally unwelcome ethnic chauvinism that results in the intra-urban violence which his *Maximum City* seeks to better understand. Historian Gyan Prakash further analyses this narrative of the city's transition from secular cosmopolitanism to intolerant nativism in the opening chapter to his *Mumbai Fables* (2010). For more on Bombay's 'descosmopolitanisation,' see Appadurai, 'Spectral Housing and Urban Cleansing' (2000).

2 In a widely reposted entry on his personal blog, *bigb.bigadda.com*, that has since been removed, Bachchan noted that 'the [*Slumdog Millionaire*] idea, authored by an Indian and conceived and cinematically put together by a westerner, gets creative globe recognition' whereas similar Indian films do not (Ramesh). Many interpreted this as a direct criticism of the film by one of India's most celebrated actors, although 'Big B', as he is popularly known in his home country, has denied that this was his intention.

3 The intense media interest in how much the young slum-dwellers chosen by Danny Boyle and his producers to play key roles in the film earned for their work prompted the director and his distributors to release public statements, which can be read in full online: http://thinkprogress.org/politics/2009/01/28/35422/slumdog-millionaire-child-actors/

4 Smita Mitra (2011) critiques some of these interventions in a recent article that suggests visiting artists' lack of lasting engagement with the slum.

5 Debjani Ganguly's *Caste and Dalit Lifeworlds* (2005) offers a useful overview of Bombay's representation in Dalit literature from the eighteenth to the late twentieth century (pp. 178–92).

6 See Hochschild (2005) and Kapur (2004) for characteristically favourable reviews of *Maximum City*, which commend Mehta's skilful handling of his potentially overwhelming subject matter.

7 See Dirks (2001) and Snodgrass (2006).

8 In his introduction to the short story anthology *Mumbai Noir* (2012), Tyrewala notes the profusion of 'overly romanticised' and 'exaggerated' portrayals of the bar line in recent Bombay literature (p. 16). By contrast, Avtar Singh's contribution to this volume, 'Pakeezah', is an effectively understated dramatisation of the sexual and emotional obsessions that the city's secretive nightlife provokes.

References

Adiga, A. (2011) *Last Man in Tower*. London: Atlantic Books.

Anand, M. R. (2009 [1935]) *Untouchable*. London: Penguin.

Appadurai, A. (2000) 'Spectral Housing and Urban Cleansing: Notes on Millennial Mumbai', *Public Culture*, vol. 12, no. 3: pp. 627–51.

Appadurai, A. (2002) 'Deep Democracy: Urban Governmentality and the Horizon of Politics', *Public Culture*, vol. 14, no. 1: pp. 21–47.

Baer, B. (2009) 'Shit Writing: Mulk Raj Anand's *Untouchable*, the Image of Gandhi, and the Progressive Writers' Association', *Modernism/Modernity*, vol. 16, no. 3: pp. 575–95.

Belliappa, M. (2008) 'Bombay Writing: Are You Experienced?' *The Antioch Review*, vol. 66, no. .2: pp. 345–62.

Boo, K. (2012) *Behind the Beautiful Forevers: Life, Death, and Hope in a Mumbai Undercity*. New York: Random House.

Chandra, V. (2007) *Sacred Games*. New York: HarperCollins.

Chakrabarty, D. (2002) *Habitations of Modernity: Essays in the Wake of Subaltern Studies*. Chicago, IL: U of Chicago P.

Clifford, J. (1988) 'On Ethnographic Authority', in *The Predicament of Culture: Twentieth-Century Ethnography, Literature, and Art*. Cambridge, MA: Harvard UP, pp. 21–54.

Company (2002) Film. Directed by Ram Gopal Varma. USA: Eros Entertainment.

Film4 (no date) 'Danny Boyle on *Slumdog Millionaire*', *Film4.com* [Online]. Available at http://www.film4.com/reviews/2008/slumdog-millionaire (Accessed 28 September 2012).

Davis, M. (2005) *Planet of Slums*. London: Verso.

Deshpande, S. (1988) *That Long Silence*. London: Virago.

Dirks, N. (2001) *Castes of Mind: Colonialism and the Making of Modern India*. Princeton, NJ: Princeton UP.

Forster, E. M. (2009 [1935]) Preface. In Anand, M. R. *Untouchable*. London: Penguin, pp. v–viii.

Ganguly, D. (2005) *Caste, Colonialism and Counter-Modernity: Notes on a Postcolonial Hermeneutics of Caste*. London: Routledge.

Hansen, T. B. and Verkaaik, O. (2009) 'Urban Charisma: On Everyday Mythologies in the City', *Critique of Anthropology*, vol. 29, no. 1: pp. 5–26.

Hawkins, G. (2006) *The Ethics of Waste: How We Relate to Rubbish*. Lanham, MD: Rowman & Littlefield.

Hochschild, A. (2005) 'Underworld: Capturing India's Impossible City', *Harper's Magazine*, February [Online]. Available at http://harpers.org/archive/2005/02/underworld/ (Accessed 20 February 2016).

Kapur, A. (2004) '"Maximum City": Bombay Confidential', *The New York Times Book Review*, 21 November [Online] Available at http://www.nytimes.com/2004/11/21/books/review/maximum-city-bombay-confidential.html (Accessed 20 February 2016).

Mazumdar, R. (2007) *Bombay Cinema: An Archive of the City*. Minneapolis, MN: U of Minnesota P.

Mehta, S. (2004) *Maximum City: Bombay Lost and Found*. New York: Vintage.

Mitra, S. (2011) 'The Slum of all Parts', *Outlook*, 28 March [Online]. Available at http://www.outlookindia.com/magazine/story/the-slum-of-all-parts/270932 (Accessed 28 September 2012).

Naipaul, V. S. (1964) *An Area of Darkness*. London: Andre Deutsch.

Naipaul, V. S. (1976) *India: A Wounded Civilization*. New York: Vintage.

Naipaul, V. S. (1990) *India: A Million Mutinies Now*. New York: Penguin.

Nandy, A. (1998) 'Indian Popular Cinema as a Slum's Eye View of Politics', in Nandy, A. (ed.) *The Secret Politics of our Desires: Innocence, Culpability, and Indian Popular Cinema*. London: Zed Books: pp. 1–18.

Nixon, R. (1992) *London Calling: V. S. Naipaul, Postcolonial Mandarin*. Oxford: Oxford UP.

Parinda (1989) Film. Directed by Vidhu Vinod Chopra. [DVD] USA: Digital Entertainment.

Prakash, G. (2010) *Mumbai Fables*. Princeton, NJ: Princeton UP.

Rao, V. (2006) 'Risk and the City: Bombay, Mumbai and Other Theoretical Departures', *India Review*, vol. 5, no. 2: pp. 220–32.

Rao, V. (2011) 'A New Urban Type: Gangsters, Terrorists, Global Cities', *Critique of Anthropology*, vol. 31, no.1, pp. 3-20.

Roberts, G. D. (2004) *Shantaram: A Novel*. New York: St. Martin's Press.

Rushdie, S. (1991) *Imaginary Homelands*. London: Granta.

Rushdie, S. (2006 [1981]) *Midnight's Children*. London: Vintage.

Salaam Bombay! (1998) Film. Directed by Mira Nair. [DVD] USA: Cinecom.

Satya (1998) Film. Directed by Ram Gopal Varma. [DVD] USA: Digital Entertainment.

Slumdog Millionaire (2000) Film. Directed by Danny Boyle. [DVD]. UK: Pathé.

Snodgrass, J. (2006) *Casting Kings: Bards and Indian Modernity*. Oxford: Oxford UP.

Thomas, R. (2008) 'Indian Cinema: Pleasures and Popularity', in Dudrah, R. and Desai, J. (eds.) *The Bollywood Reader*. Maidenhead: McGraw-Hill: pp. 21–31.

Tyrewala, A. (2012) *Mumbai Noir*. New York: Akashic Books.

5   The stakes of waste aesthetics

Waste Matters began with the critical premise that creative representations of urban waste offer valuable tools with which to diagnose and critique the intersecting structures responsible for the intensely uneven global development of the current era. Starkly apparent at the actual and figurative margins of today's cities, the preceding chapters examine a range of texts that place discarded things, degraded spaces and devalued people at the heart of their urban imaginaries. In doing so, they bear critical witness to the social and environmental injustices that emerge from the unwelcome collusion between frequently suppressed colonial histories and new global networks of economic and political oppression.

Both compelling and instructive, this varied body of work also reveals the aestheticisation of waste to be an inescapably fraught undertaking. First, privileged artists who directly engage the margins must carefully navigate their own positions of relative power vis à vis their sidelined subjects if they are to avoid simply consolidating their own cultural authority at the expense of lending a strong voice to the disempowered. As discussed in the previous chapter, Suketu Mehta's intricate and engaging ethnography of the 'other Bombay', *Maximum City* (Mehta 2004), garners atypical attention to the city's most devalued residents and spaces. However, the inherent gender bias of the male gaze he casts on the women who dance in the 'bar line' reproduces the uneven power dynamics that maintain his subjects' sidelined status.

Transforming human and material rubbish into art also risks minimising the vast and troubling inequities that lead to its production. By asserting the inherent beauty and value of overlooked urban existence, authors and artists must be careful not to understate the deprivations from which it emerges. Alert to this potential pitfall, as discussed in Chapter 1, Patrick Chamoiseau's *Texaco* (1997) avoids idealising the necessary inventiveness of the urban margins by ironically juxtaposing the Urban Planner's naively celebratory diary excerpts alongside Marie Sophie's narrative of struggle. Doing so demonstrates how outsiders can easily overlook slum-dwellers' painful reality in favour of applauding their innovative survival skills. In Chapter 2, Chris Abani's *Graceland* (2004) further demonstrates that marginal creativity cannot be divorced from the conditions of its production in his portrayal of Elvis' embattled artistic development, which is beset by violence and prejudice.

Waste aesthetics, it seems, can offer both incisive critique and problematic conciliation. As shown in Chapter 3, the privileged gentrifiers portrayed in Dinaw Mengestu's *The Beautiful Things That Heaven Bears* (2007) dramatise how a limited aesthetic experience of urban waste can all too easily substitute *for* rather than facilitate genuine engagement *with* the margins. Concerned with shoring up individual wealth and domestic comfort, these affluent Washington, D.C. residents fetishise particular kinds of once obsolescent material – outdated technology, old furniture – at the expense of recognising the discriminatory systems of accumulation and excess that produce such waste. With little care for the underprivileged residents who have long inhabited Logan Circle, they likewise treat their physical environment as the mere backdrop for the staging of their selectively cosmopolitan lifestyles. Their reclaimed decorative discards satisfy the gentrifiers' self-serving desire to recuperate lost material potential, yet they continue to deny the true nature and extent of their relationship to the city's margins.

Building on the implicit engagement with the conciliatory potential of waste aesthetics in the literary texts discussed above, this final chapter examines two documentary films that explicitly foreground the stakes of creating art from waste. Directed by Lucy Walker, *Waste Land* (2010) provides an account of modern artist Vik Muniz' transformation of Rio de Janeiro trash pickers into mixed-media portraits. With similar attention to its complex creative process, Andrew Garrison's *Trash Dance* (2013) narrates how American choreographer Allison Orr stages a dance performance in collaboration with the garbage workers of Austin, Texas. If, as shown in the previous chapters, dominant and uneven modes of urban waste production are maintained by highly contradictory modes of seeing and deliberately *un*-seeing the ubiquitous human and material rejects that occupy the urban margins, these multi-layered visual texts offer an invaluable complement to literary representations of this paradoxical visibility.

Both creative and corrective, *Waste Land* and *Trash Dance* are what we might call innovative 'remediations' of urban waste – a term that evokes both their distinctive form and their concern with arresting environmental damage. Artworks in their own right, the films dramatise the initial conversion of urban waste into visual and performance art, revealing the inherent tensions of such an endeavour. The meta-commentary they provide exposes both the predictable and unexpected challenges faced by Muniz and Orr as they attempt to facilitate genuinely collaborative waste art projects in their respective cities. By appropriating the tasks and materials central to their participants' daily waste work, both artists defamiliarise the bodies, machines and spaces that typically conceal the far-reaching effects of excessive consumption. Their formally inventive critiques of marginal disempowerment thus engage, rather than console, their respective audiences and participants by demonstrating their mutual embeddedness in fragile urban ecosystems.

As compelling visual extensions of Muniz and Orr's artworks, the films themselves also highlight the troubling intersection of social and environmental injustices in both Rio de Janeiro and Austin. By re-casting these artistic projects in self-reflexive documentary form, Walker and Garrison caution against substituting one dominant mode of seeing the urban margins for another singular perspective.

If aestheticising waste has the potential to inure its audience to the urban problems that it creatively represents, the distinctive layered form of both *Waste Land* and *Trash Dance* refuses consolation, instead providing a suggestive analogy for responsible production and engaged consumption that asks the audience to not only recognise urban waste, but also to re-examine their own perceptions of the uneven city dynamics through which it is produced.

'Pictures of Garbage': collaborative photography

A globally renowned and highly paid photographer, Vik Muniz must overcome a striking imbalance of power when he embarks on a collaborative art project with a group of disenfranchised Brazilian waste workers. By creatively remediating this process, Lucy Walker's *Waste Land* exposes the tensions inherent in the creation of waste art, encouraging its critical consumption. Muniz himself demonstrates awareness of his own economically and socially privileged status, explicitly stating his desire to exploit his celebrity for the benefit of those less fortunate. Having left his own relatively humble beginnings in São Paulo far behind, he returns to Brazil to work with the catadores or 'pickers' of Rio de Janeiro's euphemistically named Jardim Gramacho – not a garden, but a rubbish dump that was, at the time of filming in 2008, the largest in the world in terms of the daily quantity of refuse it received. These informal workers make a living by scavenging recyclable materials from the dump, which they then sell on to commercial recycling companies. Muniz takes photographs of some of the catadores, modelled in some cases after existing works by well-known artists. He then invites them to help transform these 'Pictures of Garbage' into large mixed-media portraits using the waste materials that they work with on a daily basis. He re-photographs these collages and donates the proceeds from their sale at auction to the pickers' union to help with the purchase of equipment and resources.

Despite his apparent self-awareness, critics have remained divided as to whether Muniz successfully challenges the marginalisation of his 'wasted' subjects. Notably absent from the largely positive reviews linked on the film's official website are those by Peter Bradshaw (2011) and Roger Ebert (2011), both of whom raise the tricky question of whether the art project documented in the film is exploitative – a question to which Bradshaw responds with a qualified yes.[1] Critic Paulo Moreira likewise observes that Muniz plays 'a rather comfortable part in an uplifting tale as the redeemer of a small group of pickers', which lacks the 'collaborative effort' he finds in other recent Brazilian documentaries that more explicitly reveal and attempt to overcome the hierarchical relationship between privileged film-maker and socially marginalised protagonist (Moreira 2013, pp. 253, 258).

While Stephen Holden (2010) is perhaps somewhat unfair to describe Muniz as 'manically happy' throughout the film, his optimistic outlook does, at times, sit uneasily alongside the evident hardships experienced by the catadores. However, the textured form of Walker's documentary effectively reduces the figurative distance between this celebrity artist and his disenfranchised collaborators by including their respective backstories in the form of voice-overs and individual interviews. As critic John Parham suggests, Muniz' first-hand account of his own limited resources and options during his 'lower middle-class' upbringing in São Paulo 'forges an authentic social connection with the *catadores* which legitimates [Muniz'] involvement in the project and explains his apparent rapport with them' (Parham 2016, p. 201). The additional context provided by Walker's film suggests that the underlying motivation for Muniz' project is generous rather than self-serving given his personal experience of socio-economic disadvantage.

In addition to suggesting an affinity with a younger, less fortunate, Muniz, the catadores' individual interviews reveal their autonomous efforts to articulate their rights and demand social recognition. These insights into their personal stories reveal the dump to be what anthropologist Kathleen Millar calls 'a site of subject-making' before Muniz and his team arrive (Millar 2014, p. 45). Working mother Magna de França Santos, for example, re-enacts the social confrontations she has experienced while taking public transport home from the dump. When her fellow passengers mock or complain about the way she smells after work, she refuses to be ostracised, instead pointing out the value of her work – the alternative to which, she explains, would be prostitution. She punctures the wilful ignorance that allows other city residents to disparage workers such as herself, asking those on the bus where they think their garbage goes. Magna's short performance skilfully demonstrates the double standards on which their thoughtless dismissal of waste rests. Assuming a role that is at once assertive and educational, she refuses her social degradation and effectively adopts the moral high ground.

Whereas Magna's political activism takes place in the everyday urban space of the public bus, the film also reveals the coordinated efforts of Tião (Sebastião Carlos dos Santos), president of the Association of Pickers of Jardim Gramacho to raise the profile of the workers. An energetic and charismatic presence, he is filmed leading a protest in front of the local mayor's office, calling for the pickers' work to be recognised as an official sector of the recyclable materials industry. Tião emphatically refuses to be overlooked by the state, literally occupying a central urban location in order to articulate his demands and those of his co-workers.

Walker's inclusion of these backstories and personal interviews recognises the workers' deliberate self-presentation. In their struggle to change perceptions of their devalued labour at an individual and collective level, Magna, Tião and the other catadores demonstrate an economic and environmental sensitivity that precedes Muniz' arrival. Already socially conscious and politically active, the catadores are nevertheless affected by their artistic collaboration with Muniz. For some, like Magna, the renewed self-esteem gained through participation provides the catalyst for leaving the dump and finding less precarious employment. However, as is revealed at the end of the film, not all the catadores are able to make such changes, raising the question of whether the project has enduring benefits for the workers.

Walker's documentary does not shy away from the potential limitations of the waste aesthetics forged by Muniz' portraits and, indeed, by the film itself. In keeping with its overall self-reflexive tone, *Waste Land* includes a scene in which Muniz, his wife and fellow artist Janaina Tschäpe and his production director Fabio

Ghivelder debate the impact of the project on his collaborators. Muniz himself defends the potentially short-lived relief it might offer the catadores, most of whom have no option other than to continue working on the dump once the portraits are complete. 'You don't have to change their lives forever', he argues, claiming that simply changing their perspective on themselves and their work, even for a short while, has merit. In his later analysis of the film, art scholar Frank Möller likewise asserts the psychological and emotional benefits of creative collaboration for the catadores. Regardless of how the finished portraits are displayed and received, he notes, 'nobody can take the experience of participating in the production of works of art away from [the catadores] because this experience is ingrained indelibly in their individual and social memory' (Möller 2013, p. 123). If participation in Muniz' project has enduring benefits for the catadores – notably a stronger sense of self-worth and community – *Waste Land* also demands more from its audience than a passing appreciation of marginal beauty. By revealing this debate, the film provides an essential reminder of the paradoxical nature of waste art. While Muniz' project does indeed elicit recognition of and respect for catadores, it still risks exploiting them at the same time. By inviting the viewers to reflect on the stakes of their spectatorship, the film directs them towards critical engagement with the margins that they habitually put out of sight and mind.

Appropriating waste matter

Although Walker's film shows Muniz self-consciously reflecting on how the art project will impact his collaborators, the subsequent critical focus on his relationship with the catadores speaks to an enduring anxiety about the potentially exploitative nature of transforming waste into art, particularly when that entails the aestheticisation of profoundly disenfranchised human subjects. While clearly a valid concern, far fewer have discussed what the film reveals about the complex urban ecology in which the catadores are embedded. Kevin Corbett offers some insight into this critical tendency, placing *Waste Land* within the emergent genre of what he terms the 'post-issue/advocacy' documentary, characterised by films that have some sort of '"issue" clearly present but, at best, in the middle- or background' (Corbett 2013, p. 134). In his view, *Waste Land* seems to 'gloss over the economic and political conditions that are as much a part of these people's lives as is the garbage they work in' (Corbett 2013, p. 131). If this is the case, many responses to the film to date have arguably repeated this 'glossing over' of the material conditions in which the catadores work, focusing instead on Muniz and his celebrity artist status.

While Corbett detects a departure from sociopolitical concerns in the inventive form of *Waste Land*, the documentary might alternatively be seen to advance a new kind of filmic advocacy, more suited to the complexity of the intersecting environmental and social challenges that it addresses. The 'issue' of urban waste demands a formally innovative critique that exposes its spatial, human and material dimensions and the ways in which they overlap. As John Parham suggests, an 'interplay between modes [e.g. poetic, expository] is usually the only feasible way to represent, as the film does do, something as complex as ecological posthumanism' (Parham 2016, p. 197). An exclusive critical focus on the catadores' relationship with Muniz neglects their equally important interactions with a range of other urban actants, both human and non-human, not least the dangerous discards that they deal with on a daily basis.

The critical waste aesthetics forged by Muniz and Walker are alert to the intersecting social and environmental injustices that together devalue the catadores and the work they carry out. With regard to the former, there is undoubtedly an urgent and historic need for advocacy on behalf of the catadores. In Rio de Janeiro, the quest for the recognition of their basic humanity has long been central to the lives of the city's urban poor. In her comprehensive study of the city's favelas, Janice Perlman notes the 'ongoing multigenerational struggle to "become gente" – literally to become a person, to move from invisible to visible or from a nonentity to a respected human being' that pervades marginal urban existence (Perlman 2010, p. 7). Having been, as she puts it, so long 'exclude[d] from the category of personhood', the assertion of the dump workers' humanity by artists such as Muniz is not just a poignant affective appeal, but an essential political demand (Perlman 2010, p. 316).

However, the social exclusion and subsequent dehumanisation of Rio's catadores is inseparable from the equally discriminatory environmental ideology espoused by the city authorities. The marginalisation of this important workforce began for many long before they actually crossed the threshold into the dump. Kathleen Millar notes that many catadores were forced to relocate to Jardim Gramacho in the 1990s when the then mayor initiated a series of urban reforms intended to regulate the use of public space and highlight the city's patrimony. Explicitly referred to as 'straightening-up the house' or 'cleaning the landscape', these efforts extended beyond the mere metaphorical displacement of many of those who lived and worked on Rio's streets. With the city's informal economy under threat and scarce employment available in the formal sector, those who were forced into seeking work on the dump were literally discarded by the state, just as their new source of employment – recyclable materials – had been thrown out. By recasting these 'wasted lives' as visual art, Muniz demands that his audience newly regard those urban residents that they wilfully ignore. Yet while his poignant individual representations rehumanise the disenfranchised catadores, their distinctive form also evokes the wider environmental context for their marginalisation, reminding the audience that urban waste is a complex assemblage that cannot be reduced to a singular social issue.

Walker's film highlights Muniz' concern with exposing the varied dimensions of Rio's urban waste by showing how he literally switches between a human and non-human perspective on the dump during the creative process. During his initial research for the project, Muniz takes a number of panoramic aerial shots of Jardim Gramacho, which capture the vast extent of the waste matter the catadores work with. The Sisyphean task they face in trying to reduce it is overwhelming. Although these bird's eye views of the dump are shown in Walker's film, they do not appear in Muniz's finished photographic series. When discussing the direction of the project with his colleague Fabio, they both note that the aerial shots lack the 'human factor' that is what they term 'the best part' of the dump. Muniz literally rehumanises the perspective from which he photographs the catadores in order to make them more clearly visible to his audience. However, his creative reuse of waste materials in his composition of the portraits balances their generic anthropocentrism.

In keeping with Muniz' stated desire to 'change the lives of a group of people with the same material that they deal with every day', he redirects the catadores' salvaging skills towards the collection of discarded metal, plastic and glass objects, which they then use to recreate collages of the photographic portraits for which the workers have already posed. Placing the pickers' undervalued labour at the heart of the creative process dignifies the tasks that they regularly perform without adequate compensation, protection or recognition. Modelled after well-known paintings, these striking images render the usual proximity of the catadores to rubbish beautiful and compelling rather than shameful and disgusting. Bereaved mother Isis Rodrigues Garros, for example, assumes the fatigued stance of Picasso's 1904 'Woman Ironing'. Emblematic of working-class hardship in early twentieth-century Paris, this earlier image resonates with the physical and emotional exhaustion of Rio's pickers. By recreating this now iconic picture from discarded objects – pots, shoes, bottles, cans, dust and dirt – Muniz' portrait yokes together the original's call for social compassion with a new environmentalist demand (see Figure 5.1). By defamiliarising both

Figure 5.1 Muniz, Vik (2008) *Woman Ironing (Isis)* from *Pictures of Garbage*. Art © Vik Muniz/Licensed by VAGA, New York, NY

its artistic precedent and the waste matter from which it is formed, the portrait surprises and visually admonishes the viewer who habitually ignores the social and environmental impact of their own waste production.

Recognition of their subjects' inherent humanity is an essential element of the portraits' appeal, but their distinctive composition also suggests the porous boundaries of the human subjects they depict. Isis and her fellow pickers are not just figuratively shaped by the waste in which they work, but they are actually formed by it. Walker's documentary further highlights this in its account of the elderly Valter dos Santos who dies during filming. His optimistic catchphrases, such as '99 não é 100' ('99 is not 100') resonate around the dump after his passing, reminding the catadores to take pride in their salvaging work, no matter how much Rio's wealthier and more powerful residents may belittle their contributions to preserving the city's ecosystem. The film's coda attributes Valter's death to lung cancer, the implication being that his years spent inhaling toxins on the dump have caused his disease. By exposing the cellular intimacy between the catadores and their polluted surroundings, *Waste Land* refuses a superficial aesthetics. In keeping with what Stacy Alaimo (2010) identifies as the 'trans-corporeality' of the human body, the way in which the decomposing organic and plastic discards impact and alter Valter's physical health suggests the profound interconnections between Jardim Gramacho's various human and material constituents.

The unwelcome intimacy between the catadores and their toxic environments is further highlighted by Walker's insertion of disturbing wordless montages into the film's narrative trajectory, which reveal the full extent of the perilous conditions in which the catadores work. Accompanied by a haunting score composed by Moby, these scenes evoke the uncanny nature of the dump, the existence of which is consciously repressed by the majority of the city's residents. Perhaps most troubling are the night-time shots, showing open fires that suggest the profoundly other- or under-worldly nature of this setting (see Figure 5.2). These unnerving scenes not only bear important visual witness to the dump, but they also invite the audience to acknowledge their own role in its production. The film thus produces a form of cognitive dissonance in the viewer who is accustomed to denying the existence of that which they must now confront. In this way, the film serves as a productive form of visual disturbance, which prompts conscious reflection on the inequities of the waste system that it documents.

Waste aesthetics on display

On completion, the finished portraits are sold in London's high-end Phillips de Pury auction house, the sanitised nature of which offers a striking contrast to the toxic site of their production. The literal and figurative distance travelled by Muniz' 'Pictures of Garbage' suggests the significant role that such art might play in exposing the unseen destination of unthinking consumption to a powerful global audience. By putting the finished portraits into commercial circulation, Muniz offers a knowing critique of the exclusive world of 'high art'. The images command an exchange value which far exceeds that of the recyclable materials

Figure 5.2 Jardim Gramacho at Night (2010) from *Waste Land.* Dir. Lucy Walker

from which they are made. Their sale affirms Michael Thompson's (1979) 'rubbish theory', which suggests that the manner in which those with significant wealth and associated power assign value to discarded objects is a means of social control. The shifting value of rubbish, which is deliberately foregrounded by Muniz, demonstrates the highly subjective nature of this process. By turning the portraits into pragmatic tools of wealth redistribution, Muniz successfully subverts this mechanism to provide tangible help for the catadores.

Nevertheless, the profit-driven cultural economy ultimately remains intact. Those concerned about the effect of Muniz' project on the catadores may well ask whether the critical irony of their humble origins is lost on those who purchase the portraits at great expense for their personal art collections. By selling the works directly to auction, they are not initially displayed in public galleries. They immediately enter an elite world of art collection that rests on the same principle of private acquisition that subtends the dominant waste regime. Watching the auction take place, including phone bids from potential buyers who have never seen the portraits in person, a viewer might think that those with the money and inclination to purchase them are investing in a refined, contained form of urban waste, albeit one that does test the accepted definition of what constitutes 'high art'. Furthermore, as the film shows, Tião accompanies Muniz to the sale of his portrait. His emotional reaction to the winning bid of £28,000 poignantly reminds the viewer of the gaping inequity between those who can afford such luxuries and those for whom they will be forever unattainable. It is to Tião's credit, in fact, that on his return to Rio, he gracefully expresses a newfound appreciation for modern art rather than insurmountable anger at the injustice of his social marginalisation.

Despite their collaborative production and innovative composition, the circulation of Muniz' portraits as highly valued cultural artefacts still risks screening their elite audience from the seriousness of the social and environmental injustices that constitute Jardim Gramacho. By highlighting other ways in which the portraits are displayed, *Waste Land* crucially remediates this most precarious aspect of Muniz' venture. The film shows how they are publically exhibited at a later date in Rio's Museum of Modern Art as part of a retrospective of Muniz' work. Although Muniz himself is the focal point of the exhibit, the featured catadores attend the opening night, sharing in the public admiration for both their images and their waste work. In a brief but significant moment of direct recognition, newspaper photographers replace Muniz, taking the catadores' pictures in order to report their achievements within their home city and further afield. Additionally, towards the end of the film, Muniz is shown taking smaller-scale portraits into the catadores' homes where he hangs them in whatever space is available. In striking contrast to the gleaming white interior of the Phillips auction house, these basic domestic galleries highlight the different values embedded in the portraits. While there is no commercial market for them here, they bequeath a vast amount of self-esteem and dignity to the catadores they feature.

Lastly, *Waste Land* itself further disseminates the catadores' images and their backstories to a wide global audience. While critics such as Kevin Corbett (2013) have found the film's social engagement to be somewhat understated, as mentioned above, the wide and continued distribution of the documentary serves to raise awareness of Rio's extensive urban waste and the problematic wider context for its proliferation. Highlighting the specific plight of the catadores has become even more important with the closure of Jardim Gramacho in 2012 after thirty-four years of continuous use, which has led to displacement and loss of income for many of the workers. By depicting the varied ways in which the portraits circulate amongst disparate audiences, *Waste Land* extends Muniz' own commentary on the need for responsible production and engaged consumption of waste aesthetics. Just as the 'Pictures of Garbage' elicit multiple interpretations and valuations, so, too, does the complex challenge of urban waste demand consideration from diverse perspectives. The layered form of *Waste Land* allows the film's audience to rehearse these different viewpoints, enabling them to better understand and productively adapt their own interpretations not only of waste art, but crucially the complex assemblage of discarded things, people and places from which it emerges.

The Trash Project: collaborative choreography

Like Muniz, choreographer Allison Orr is committed to dignifying waste work through aesthetic transformation, a challenging creative process well captured by Andrew Garrison's *Trash Dance*, which documents her collaboration with the employees of Austin's Solid Waste Services Department (since renamed Resource Recovery). The film follows Orr as she shadows the city's various clean-up crews over a period of eight months in 2009, trying her hand at different duties from litter picking to street sweeping and dead animal collection. This experiential research culminates in her choreography of a 75-minute modern dance piece featuring twenty-four sanitation workers and the equipment they use, including sixteen large vehicles. Following intense rehearsals, it is performed on a decommissioned airport runway in the city to an audience of several thousand with a live score composed and played by Graham Reynolds (see Figure 5.3). In response to public demand, two encore performances took place in 2011 featuring most of the show's original cast (Seitz 2011).

Orr's willingness to carry out waste work herself is essential to forging a non-hierarchical relationship with the project participants. Whereas Muniz visits Jardim Gramacho and observes Rio's catadores at work, his research does not extend to picking recyclables himself. Orr, however, literally gets her hands dirty. By deferring to the workers' professional expertise when she is struggling with a new or unfamiliar task, she successfully offsets her own status as the project's creative director. She further reallocates her own authority by inviting the project participants to share any individual performance skills they have, discovering a talented rapper, an accomplished harmonica player and a competitive skate dancer amongst them. She includes these additional skills in the final dance, giving space to the workers' self-expression alongside the coordinated sequences that she directs. The result is a truly collaborative piece that showcases the performers' wide range of professional and personal abilities.

The relative privileges that Orr must negotiate vis à vis Austin's waste workers are both obvious and implicit. As an educated white American, economic and racial prejudices generally work in her favour, unlike the intersecting social disadvantages experienced by the largely African American and Latino sanitation workforce. In addition to gaining first-hand experience of their jobs, Orr overcomes potential mistrust from the workers by speaking Spanish where relevant and making a specific separate appeal for participation from the female employees who are a minority within an already marginalised team. Her overtures are initially greeted with bemusement. Recycling collector Lee Houston explains, somewhat tongue-in-cheek, that 'it was weird, you know, for a little short-haired white lady [to be] interrogating people'. For their part, however, the workers respect Orr and warmly accept her into their team, laughing at her occasional naivety without dismissing the project that she is trying to instigate.

Underlying the workplace banter to which Orr gamely responds is the suggestion that her privileged social standing entitles her to environmentalist sensibilities that her collaborators cannot afford. 'Are they organic?' one worker jokes when Orr brings in doughnuts to share at a staff meeting, 'Are they green?' However, it is, in part, the stereotypically complacent environmentalism of Austin's liberal white majority – their avoidance of dietary toxins, their willingness to recycle, their appreciation of apparently clean, open spaces – that Orr seeks to unsettle through the literal spectacle of a 'trash dance'. Her choreography delivers a socially conscious environmentalist message that asserts the invaluable role Austin's waste workers play in preserving the city's delicate ecosystem. In keeping with ecocritic Heather Sullivan's defence of 'dirty aesthetics', Orr's project further suggests that care for the environment must attend to precisely those places, things and people that persistently disrupt the desired 'purity' of our material surroundings (Sullivan 2012, p. 515). Unlike Rio's catadores, who strategically assert their indispensable role in maintaining urban sustainability, Austin's sanitation workers do not see themselves as active environmentalists. In addition to making their work visible and impressive to a wide range of Austin residents, Orr encourages them to appreciate the social and environmental impact of the service they provide, which maintains the smooth operation of the city and the comfort of its residents.

Incorporating waste machines

Despite the differences between Orr and her collaborators, which reflect disparities in the city as a whole, she strives to create what she calls a 'shared moment' between herself, the dancers and the audience that demonstrates their interconnectedness via the actual and symbolic medium of urban waste. Her desire to make Austin's waste workers clearly visible resonates with the aims of Muniz's 'Pictures of Garbage', as does her attention to individualising this under-appreciated workforce. Interestingly, both Orr's project and Garrison's complementary remediation extend this desired urban collectivity to include the nonhuman. Although Orr's primary goal is to spotlight the people who facilitate Austin's waste collection and disposal, her project also highlights a variety of other actants within this urban ecosystem, notably the often cumbersome machines with which the sanitation crews work. Like *Waste Land*, *Trash Dance* suggests the permeable boundaries of the waste workers' bodies. Yet while Muniz drew attention to the unwelcome cellular mutations caused by the toxins emanating from Jardim Gramacho, Orr suggests the productive extension of the Austin workers' physical capabilities through their use of trucks and equipment that serve as mechanised prosthetics. Watching both Orr's performance and Garrison's film does not just establish a broader sense of urban community, but it also recalibrates notions of exactly what such a collectivity might entail.

At first, Orr's choreography deliberately draws attention to the labouring bodies of the waste workers themselves. When discussing the rationale for her project, she explains, 'for me, it all starts with a guy picking up the trash'. This is mirrored in the simple opening sequence of the performance, which shows a single worker walking on stage and slowly filling a black bag with unseen rubbish. In interviews with the crews, Garrison discovers that all of them have second jobs that they go to on their 'days off'. If conspicuous consumption now forms the basis of leisure time in the US, Austin's waste workers are excluded from this social ritual, only coming into contact with the inevitable material excess it produces. They have scarce free time, spending their days performing indispensable functions and services. Orr's choreography subverts the workers' social casting as automatons, instead foregrounding the physical elegance of the work they carry out. By incorporating the sanitation crews' professional skills into the final performance – showing them rapidly lifting and loading rubbish bins in perfect sync, for example – Orr makes explicit the inherent beauty of the daily activities that she has witnessed.

Orr further defamiliarises the abstracted labour of the waste workers by drawing attention to the machines that facilitate their different tasks, from the large vehicles they drive to the simple 'pick sticks' that allow them to grasp discarded litter without constantly having to bend down. Throughout the documentary, montages of different trash crews at work emphasise the synchronicity between the waste workers and this commonplace waste technology. During her research, Orr is excited to watch crew member Gerald Watson demonstrate how he efficiently positions and operates his bucket truck in order to collect street sweepings: 'That's dancing right there', she notes, 'you're coordinating space, time and energy'. By directing the audience's attention to such procedures, Orr's choreography reveals and dignifies the waste work that is dismissed as mundane by the urban majority. Her transformation of such tasks into performance art is not only visually appealing, it is also emotionally engaging because it nurtures mutual respect for the crews' work and legitimate pride in their occupation. The incorporation of the crew's equipment into the dance further suggests the complexity of the urban ecosystem in which the audience and the performers are collectively enmeshed. Orr's waste aesthetics foreground the wide-ranging interactions between humans, machines and diverse matter that enable and sustain Austin, suggesting the need for an expansive understanding of urban community that is not wholly anthropocentric.

Particularly moving is the crane solo that forms the climax of the live performance. Throughout the documentary, Garrison films the crane's operator Don Anderson, who is one of the first to volunteer to participate in Orr's show. He articulates pride in his work, asserting that the waste workers 'are not just some dirty people who pick up trash. There is some grace to what we do'. He deftly demonstrates the crane's fluid and elegant range of movement to Reynolds' accompanying musical performance. It rises and falls, opening and shutting its metal bucket in wing-like gestures reminiscent of its avian namesake. Here, a machine that is rarely appreciated for its beauty is rendered uncannily enchanting, with Reynolds' poignant piano score deepening the audience's affective response to its mechanised display. Watching Anderson's solo elicits admiration for his skilful operation of the crane, which reveals an unexpected tenderness to his work. While neither he nor his colleagues initially see themselves as 'green', the dance situates them as invaluable custodians of Austin's urban ecosystem. Moreover, Orr's waste aesthetics here suggest that care for the environment depends less on a privileged affinity with one's physical surroundings, and more on willing collaboration with the various human, material and spatial agents by which they are constituted.

Waste aesthetics in performance

By showcasing the beauty of the sanitation crews' everyday labour, Orr succeeds in dignifying those she calls Austin's 'unsung heroes'. However, the process of aesthetic transformation inevitably diminishes some of the hazards daily encountered by the waste workers. Whereas the vast amount of material waste exposed in *Waste Land* is cautionary, Orr deals with apparently more manageable quantities of rubbish. Austin's main landfill is only shown briefly from afar in the film and actual trash is largely absent from the dance performance itself, the emphasis being on the physicality of the workers and the machines they operate.

Figure 5.3 Performers and Choreographer Allison Orr Take a Bow (2012) from *Trash Dance*. Dir. Andrew Garrison. Courtesy of trashdancemovie.com

If the affective and visual appeal of the dance risks inuring its audience to urban waste rather than demanding an overhaul of the system that produces it, the public, open-air nature of the performance provides an essential counter-point to this potentially conciliatory quality. Performed on a decommissioned airport runway, the audience must themselves enter a marginal urban space in order to watch it. Orr reclaims this defunct transport hub to showcase the undervalued workforce who themselves form one of the city's most indispensable human infrastructures. Further, the intense rainfall during the performance itself provides the audience, most of whom are sat huddled under umbrellas on the wet concrete, with a first-hand experience of the physical discomfort regularly experienced by the waste workers who must complete their rounds regardless of the elements. In this way, the performance does not only enable the city residents to see and appreciate the sanitation crews, but it also forges a more enduring experiential bond between them.

In addition to the immediate local impact of the initial dance performance, Garrison's film has productively extended its influence on perceptions of waste work. Following a private viewing by the City Manager and a number of Division Heads, the sanitation crews received a 5% pay rise.[2] Complementary international screenings at numerous film festivals have further widened the public reception of Orr's project.[3] Just as Walker's remediation ensured that Muniz' 'Pictures of Garbage' circulated beyond the elite world of contemporary art collection, *Trash Dance* enables Orr's waste aesthetics to resonate globally as well as locally, suggesting the far-reaching critical potential of such collaborative art.

Conclusion

The stakes of creating art from waste are high. Aestheticisation risks exploiting, idealising and trivialising the extreme challenges of marginal urban existence, where social disempowerment intersects with environmental degradation. Both *Waste Land* and *Trash Dance* self-reflexively acknowledge these risks in their creative remediations of two contrasting art projects. By highlighting how Muniz and Orr eschew artistic authority in favour of genuine collaboration with the waste workers of Rio de Janeiro and Austin, the films reveal a responsible and inclusive mode of production that offers an instructive contrast to the individualistic system of acquisition and excess that prevails in both cities.

By showing how their devalued labour helps to sustain the operation of their respective cities, the films elicit the audience's recognition of the waste workers' contributions. However, both Muniz and Orr eschew a wholly anthropocentric vision of urban community. In *Waste Land*, we see the unwelcome interchange between the catadores' bodies and the discarded matter with which they work. *Trash Dance*, on the other hand, embraces the flexible boundaries of the human body to showcase the elegant physicality of the waste workers' daily interactions with seemingly mundane waste technology. In both cases, the creative transformation of urban waste into visual and performance art suggests the way in which multiple agents, both human and non-human, shape and transform urban existence. They offer an environmentalist vision that calls for full attention to all that the urban ecosystem encompasses.

The local display of these artworks forges an instructive experiential bond between the featured waste workers and their audiences. Whether looking at Muniz' portraits or watching Orr's dance, the viewer is invited to recalibrate their perception of an often suppressed aspect of urban existence. The projects' immediate impact is productively extended by the documentaries' global circulation, which have attracted broad international audiences at film festivals, public screenings and online. The films' continued popularity speaks to the enduring potential of waste aesthetics to engage diverse audiences in questions of social and environmental justice despite the representational risks entailed.

Notes

1   See *www.wastelandmovie.com*
2   Thank you to director Andrew Garrison for bringing this important outcome to my attention.
3   The film also remains readily accessible online: www.trashdancemovie.com

References

Abani, C. (2004) *GraceLand*. New York: Picador.
Alaimo, S. (2010) *Bodily Natures: Science, Environment, and the Material Self.* Bloomington, IN: Indiana UP.
Bradshaw, P. (2011) '*Waste Land*', *The Guardian*, 24 February [Online]. Available at http://www.theguardian.com/film/2011/feb/24/waste-land-review (Accessed 17 February 2016).
Chamoiseau, P. (1997 [1992]) *Texaco* (trans. from French and Creole by R-M Réjouis and V. Vinokurov), New York: Vintage.
Corbett, K. (2013) '"Gleaners" and "Waste": The Post-Issue/Advocacy Documentary', *Journal of Popular Film and Television*, vol. 41, no. 3: pp. 128–35.

Ebert, R. (2011) 'Waste Land', *RogerEbert.com*, 11 February [Online]. Available at http://www.rogerebert.com/reviews/waste-land-2011 (Accessed 17 February 2016).

Holden, S. (2010) 'From a Universe of Trash, Recycling Art and Hope', *New York Times*, 28 October [Online]. Available at http://www.nytimes.com/2010/10/29/movies/29waste.html (Accessed 17 February 2016).

Mehta, S. (2004) *Maximum City: Bombay Lost and Found*. New York: Vintage.

Mengestu, D. (2007) *The Beautiful Things That Heaven Bears*. New York: Riverhead.

Millar, K. (2014) 'The Precarious Present: Wageless Labor and Disrupted Life in Rio de Janeiro, Brazil', *Cultural Anthropology*, vol. 29, no. 1: 32–53.

Möller, F. (2013) *Visual Peace: Images, Spectatorship and the Politics of Violence*. Basingstoke, Palgrave Macmillan.

Moreira, P. (2013) 'Roberto Berliner's *Born to Be Blind* and Marco Prado's *Estamira*: In Search of a New Ethics with the Other', Studies in Documentary Film, vol. 7, no. 3: pp. 249–61.

Parham, J. (2016) *Green Media and Popular Culture: An Introduction*. Basingstoke, Palgrave Macmillan.

Perlman, J. (2010) *Favela: Four Decades of Living on the Edge in Rio de Janeiro*. Oxford: Oxford UP.

Seitz, J. (2011) '"Trash", Take Two', *The Austin Chronicle*, 26 August [Online]. Available at http://www.austinchronicle.com/arts/2011-08-26/trash-take-two/ (Accessed 17 February 2016).

Sullivan, H. (2012) 'Dirt Theory and Material Ecocriticism', *Interdisciplinary Studies in Literature and the Environment*, vol. 19, no. 3: pp. 515–31.

Thompson, M. (1979) *Rubbish Theory: The Creation and Destruction of Value*. Oxford: Oxford UP.

Trash Dance (2013) Film. Directed by Andrew Garrison. [DVD]. USA: PBS International.

Waste Land (2010) Film. Directed by Lucy Walker. [DVD]. USA: Arthouse.

Conclusion

Today's cities are exemplary sites of uneven development. Characterised by stark disparities in wealth, power and wellbeing, their inherent fragmentation disavows their social and economic promise. Although a worldwide phenomenon – one need only look to the contrasting homes and incomes of London's Tower Hamlets for evidence in the global North – urban inequality is disproportionately felt in the developing cities of the global South, where long histories of colonial exploitation are inflected by new forms of discrimination, violence and neglect.

Waste Matters has shown how global authors and artists offer a vital interrogation of contemporary inequality by placing degraded spaces, devalued people and discarded things at the centre of their urban imaginaries. In their particular attention to urban waste, Patrick Chamoiseau, Chris Abani, Dinaw Mengestu, Suketu Mehta, Vik Muniz and others foreground the troubling intersection of social and environmental injustices across diverse cities. Neither wholly dystopian nor blindly idealistic, they identify new sites of collaboration and belonging while sharply critiquing the material and discursive obstacles to the creation of a truly inclusive urban politics.

There is a danger, architect David T. Fortin suggests, that by lending descriptive form to the urban margins, 'slum fictions' consolidate these dynamic and ambiguous places as concrete 'site[s] for intervention', which can therefore be managed in isolation from the complex socio-economic networks that connect them to other urban and rural spaces both nearby and distant (Fortin 2010, n.p.). Indeed, the stakes of aestheticising urban waste are high. As discussed in Chapter 5, transforming the literal and figurative margins into art risks understating the troubling collusion of economic interests, environmental disregard and social discrimination by which they are constituted. However, the texts examined throughout *Waste Matters* create, following Lefebvre (1991), innovative 'representational spaces' that are attentive to the necessarily comparative nature of urban existence. Far from portraying the precarious margins as sites of irreducible difference, their thematic and formal engagement with urban waste reveals their tangible and unseen linkages to numerous other places. Contrary to their putative lack of value, *Waste Matters* reveals human and material discards to be potent sources of historical, geographical and conceptual connection.

Histories of waste

Waste's material ubiquity suggests an attractive universality, which is reflected in cultural theories of the things we throw away. In keeping with Kevin Lynch's expansive definition of waste as 'any used thing' (Lynch 1990, p. xi), Brian Thill elegantly suggests that 'waste is every object, plus time' (Thill 2015, p. 8). Both point to the apparent inevitability of disposal. Here, the ceaseless production of waste is a seemingly indiscriminate practice that reveals a common desire for novelty.

However, the urban waste examined throughout *Waste Matters* cannot be traced to the simple passing of 'homogenous, empty time' (Benjamin 1968, p. 261). The domestic litter, industrial remnants and invisible toxins that proliferate at the margins are storied objects whose stubborn excess belies the efficiency of contemporary global capitalism. If, as Michael Thompson notes, 'the boundary between rubbish and non-rubbish moves in response to social pressures' (Thompson 1979, p. 11), the power to exert such pressures is tightly controlled in the current era. The acquisitive interests of powerful multinationals, self-serving governments and wealthy urban residents deny opportunities to the disenfranchised humans who inhabit the margins. The production of these 'wasted lives' together with the degradation of their surroundings is tied to strategic decision-making, designed to protect existing wealth and influence.

Patrick Chamoiseau's *Texaco*, discussed in Chapter 1, reasserts the historiographical potential of urban waste. A fictionalised account of an actual place, the novel describes the emergence of the Texaco slum, which battles for a foothold in the hills above Martinique's capital city of Fort-de-France. In order to survive, the slum-dwellers inventively construct homes from discarded materials they find in the streets and at the municipal rubbish dump. These objects disclose the slum's long transnational history, which has been sidelined by colonial discourse and further suppressed by neocolonial sponsorship of the island's industrialisation. Some find artefacts imported from other parts of the erstwhile French empire, others use discarded materials that reflect the frivolous expenditure of the local elite. These wasted things refuse assimilation, providing resonant material evidence of imperial investments and the disparate fortunes of the island's inhabitants. Evoking the plantation thrift of her predecessors, the protagonist Marie Sophie turns her hard-won gleaning skills to the curation of her parents' memories, further demonstrating the historical insights that an engagement with waste might provide. By compiling his slum history from diverse narrative perspectives and styles, Chamoiseau creatively appropriates this salvage methodology.

The conclusion of *Texaco* sees the slum's incorporation into the city 'proper' through an organised upgrading programme instigated by Martinique's civic authorities. Such attempts to rationalise the urban margins evoke additional histories of colonial urban planning and segregation. As discussed in Chapter 2, Chris Abani's *Graceland* (2004) portrays the violent elimination of the embattled Maroko slum as the unwelcome alternative to this kind of improvement. Both symptom and symbol of Nigeria's nervous post-independence condition, the slum clearance reflects how the state's inheritance of colonial ideals of urban order dangerously intersects with the paranoid consolidation of its own power.

In *Maximum City* (2004), Suketu Mehta also observes the inadequate urban infrastructures bequeathed to Bombay by British colonial planners, exemplified by the city's often irregular and unsanitary water supply. Modern redevelopment projects remain out of step with the lived experience of that same urban space. The many new apartment blocks that rise up from land once occupied by Bombay's textile mills suggest to Mehta a 'mediocre imagination' at work, which conceives inappropriate architecture at a remove with little care for the constraints of the specific environment nor the needs of the millions who once worked there (Mehta 2004, p. 125). Haunted by colonial occupation and industrial decline, this palimpsestic 'wasteland' shows a troubling historical continuity between past European impositions and imitative redevelopment initiatives.

The texts examined throughout *Waste Matters* cultivate a more promising reimagination of waste's manifold existing forms by putting a crucial historical counter-discourse into circulation. By tracing contemporary urban challenges such as migration, overcrowding, inadequate resources and political alienation to the enduring impacts of empire, they refuse the naturalisation of present-day poverty.

Geographies of waste

The discrepant colonial and neocolonial histories evoked by urban waste also point to its broad geographical reach. Always an immediate local challenge, the accumulation of rejected bodies and objects at the margins of developing cities suggests an enduring imbalance of power between the global North and South. The thriving international trade in hazardous waste – both legal and illegal – shows the readiness of industrialised nations to export the poisonous by-products of industry, agriculture and commerce to their poorer counterparts (Clapp 2001, Pellow 2007). Dutch-based multinational Trafigura's illegal dumping of toxic chemicals in Abidjan in 2006, which caused injuries to over 30,000 of the Ivorian port city's residents, is just one notorious case of the dangerous global mobility of unwanted matter (Leigh 2009).

Yet, while urban waste is a concrete marker of global disparity, the texts examined throughout *Waste Matters* repeatedly invoke its liminality in order to dispute the presumed geopolitical difference on which that inequality rests. The displaced figures that focalise their urban narratives repeatedly demonstrate the interconnections between the putatively distinct poles of the industrialised North and developing South. In Mengestu's T*he Beautiful Things that Heaven Bears*, for example, Sepha's exilic perspective foregrounds similarities between Washington, D.C. and his home city of Addis Ababa. From his marginal vantage point, he constructs, following Jameson (1988), a 'cognitive map' that reconfigures the relationship between the American and Ethiopian capitals to suggest their mutual imbrication in a global capitalist system that fosters inequity not only between, but also within both cities.

In bringing together a wide range of texts, *Waste Matters* models a transnational methodology that builds on their individual critiques of globalisation. Focused on disparate cities, these works' common engagement with questions of social exclusion and urban degradation insists on comparative analysis. Urban simultaneity – the experience of actually and imaginatively living in two or more cities at once – is a recurrent feature of these texts. While the aggressive policing of national borders asserts their continued material and figurative significance, the interlinked cities depicted throughout *Waste Matters* reframe the global hierarchy on which such provincialism rests by bringing diverse cities into a shared analytical frame.

Scales of waste

If examining urban waste offers urgent critical purchase on the macro-structures that create and perpetuate contemporary inequity, it is also uniquely placed to reveal the micro-effects of global capitalism. Throughout *Waste Matters*, social and environmental injustices materialise in the piecemeal housing, unstable terrain and vulnerable bodies that constitute the urban margins. While its massive scale tempts abstraction, the texts examined insistently point to the concrete lived experience of uneven development.

In Chapter 5, visual and performance art by Vik Muniz and Allison Orr highlighted the porous boundaries of the human form. The composite portraits featured in *Waste Land* display the literal constitution of Rio de Janeiro's disenfranchised catadores by the toxic matter in which they work. In *Trash Dance*, the unwieldy machines used by Austin's sanitation crews are recast as surprisingly elegant prostheses that aid their daily tasks. Whether expressing an unwelcome cellular intimacy or a productive physical modification, both projects draw attention to the numerous human and non-human actants that constitute today's precarious urban ecosystems. In keeping with recent ecocritical assertions of the agentic capacity of all matter, they resist an anthropocentric reading of the urban margins to suggest instead the ways in which diverse discards and the tools by which they are managed exert unpredictable influences and effects on the humans who share their outcast status.

In presenting urban waste as a dynamic aggregate of human and material rejects, *Waste Matters* has sought to balance social critique with environmentalist concerns. There is no doubt that asserting the fundamental humanity of those 'wasted lives' who populate the margins remains an urgent undertaking. While the texts examined repeatedly assert the creativity of such impoverished communities, the fundamental injustice of their struggle remains obscene. The millions who live and breathe inequity in cities such as Fort-de-France, Lagos and Bombay qualify appreciation for the 'vital force' inherent in all matter given the deadly agency of their polluted surroundings (Bennett 2010, p. 24).

However, *Waste Matters* resists the wholesale parsing of urban waste into its separate human, material and spatial constituents. The interconnectedness of social and environmental injustices has been a persistent refrain throughout this book. By describing and dramatising struggling slums, metropolitan rubbish dumps and gentrifying neighbourhoods, the texts examined provide a window onto exactly how and why the mutual degradation of these sites and their inhabitants occurs.

Creative representations of urban waste are thus inherently generative. By foregrounding that which has been rejected, sidelined and suppressed, contemporary authors and artists advance a persuasive critique of unacceptable disparities in urban wealth and wellbeing. In both theme and form, the texts examined throughout *Waste Matters* reveal manifold historical, geographical and conceptual connections that refute marginal alterity. In doing so, they foster a compelling and inclusive urban imaginary that urgently challenges narrow perceptions of who and what matters in today's cities.

References

Abani, C. (2004) *GraceLand*. New York: Picador.

Benjamin, W. (1968) *Illuminations* (trans. H. Arendt), New York: Shocken Books.

Bennett, J. (2010) *Vibrant Matter: a Political Ecology of Things*. Durham, NC: Duke UP.

Clapp, J. (2001) *Toxic Exports: The Transfer of Hazardous Waste from Rich to Poor Countries*. Ithaca, NY: Cornell UP.

Fortin, D. (2010) 'Slum Fictions: De-delimiting Place in Nairobi', The American Institute of Architects [Online]. Available at http://www.aia.org/aiaucmp/groups/aia/documents/pdf/aiab087177.pdf (Accessed 16 February 2016).

Jameson, F. (1988) 'Cognitive Mapping', in Nelson, C. and Grossberg, L. (eds.) *Marxism and the Interpretation of Culture*. Chicago, IL: U of Illinois P: pp. 347–60.

Lefebvre, H. (1991 [1974]) *The Production of Space* (trans. from French by D. Nicholson-Smith). Oxford: Blackwell.

Leigh, D. (2009) 'Revealed: Trafigura–commissioned Report into Dumped Toxic Waste', *The Guardian*, 17 October [Online]. Available at http://www.theguardian.com/world/2009/oct/17/trafigura-minton-report-revealed (Accessed 16 February 2016).

Lynch, K. (1990) *Wasting Away*. San Francisco, CA: Sierra Club Books.

Mehta, S. (2004) *Maximum City: Bombay Lost and Found*. New York: Vintage.

Pellow, D. N. (2007) *Resisting Global Toxics: Transnational Movements for Environmental Justice*. Cambridge, MA: MIT Press.

Thill, B. (2015) *Waste*. New York: Bloomsbury.

Thompson, M. (1979) *Rubbish Theory: The Creation and Destruction of Value*. Oxford: Oxford UP.

Index

 **Taylor & Francis eBooks**

Helping you to choose the right eBooks for your Library

Add Routledge titles to your library's digital collection today. Taylor and Francis ebooks contains over 50,000 titles in the Humanities, Social Sciences, Behavioural Sciences, Built Environment and Law.

Choose from a range of subject packages or create your own!

Benefits for you

» Free MARC records
» COUNTER-compliant usage statistics
» Flexible purchase and pricing options
» All titles DRM-free.

Benefits for your user

» Off-site, anytime access via Athens or referring URL
» Print or copy pages or chapters
» Full content search
» Bookmark, highlight and annotate text
» Access to thousands of pages of quality research at the click of a button.

eCollections – Choose from over 30 subject eCollections, including:

Archaeology	Language Learning
Architecture	Law
Asian Studies	Literature
Business & Management	Media & Communication
Classical Studies	Middle East Studies
Construction	Music
Creative & Media Arts	Philosophy
Criminology & Criminal Justice	Planning
Economics	Politics
Education	Psychology & Mental Health
Energy	Religion
Engineering	Security
English Language & Linguistics	Social Work
Environment & Sustainability	Sociology
Geography	Sport
Health Studies	Theatre & Performance
History	Tourism, Hospitality & Events

For more information, pricing enquiries or to order a free trial, please contact your local sales team:
www.tandfebooks.com/page/sales

For Product Safety Concerns and Information please contact our EU
representative GPSR@taylorandfrancis.com
Taylor & Francis Verlag GmbH, Kaufingerstraße 24, 80331 München, Germany

www.ingramcontent.com/pod-product-compliance
Ingram Content Group UK Ltd.
Pitfield, Milton Keynes, MK11 3LW, UK
UKHW021609240425
457818UK00018B/468